AF540893

Handling and Preservation of Fruits and Vegetables by Combined Methods for Rural Areas

Technical Manual

Gustavo V. Barbosa-Cánovas
Juan J. Fernández-Molina, Stella M. Alzamora
Maria S. Tapia, Aurelio López-Malo
Jorge Welti Chanes

Reprinted by arrangement with the
Food and Agriculture Organization of the United Nations

2007
Daya Publishing House
A Division of
Astral International Pvt Ltd
New Delhi - 110 002

"This book was originally published by the Food and Agriculture Organization of the United Nations (FAO) as Vaccine Manual, FAO Animal Production and Health Series No. 35".

First Indian Edition, 2007

ISBN: 978-81-70355-05-2

Published by : **Daya Publishing House**
A Division of
Astral International Pvt Ltd
4736/23, Ansari Road
Darya Ganj, New Delhi - 110002
Phone: 011-43549197
Email: info@astralint.com
Web: www.astralint.com

Printed at : **Replika Press Pvt. Ltd.**

FOREWORD

Fruits and vegetables are nutritious, valuable foods full of flavour. However, in the low-income countries, poor care and handling of these crops frequently results in loss of quality, especially when not consumed immediately. In these countries, people are not sufficiently informed on how to make technical choices for better preservation of fruits and vegetables. This manual on handling and preservation of fruits and vegetables by combined methods has been prepared in response to needs, both real and perceived, that surplus crop can be used.

The manual is the result of contributions from a selection of different authors, mainly from countries in Latin America. It contains basic concepts and operations of processing, which are essential for a better understanding and comprehensive approach to the application of the combined methods technology. Some practical examples are described step by step, including calculations and procedures required to set up this technology elsewhere. Likewise, it includes examples of modern processing techniques required to meet the high standards of quality and hygiene for food production.

This manual is divided into five chapters. Chapter one presents a global overview on trading in fruits and vegetables, it shows trends in consumption and considers some of the socio-economic issues involved in the context of post harvest food losses especially during processing and storage. Chapter two describes some concepts of harvesting and post harvest handling, storage and pest control. Chapter three focuses on the importance of the concept of water activities (a_W), and their role in food preservation. Similarly, it describes the concept of intermediate moisture foods (IMF) and the combined methods preservation technology for fruits and vegetables. Chapter four is mostly concerned with fruits, and describes the extension of the intermediate moisture concept to products containing high moisture. The chapter includes the main preliminary operations and formulations. This includes packaging, transport, storage, use of fruits preserved by combined methods and quality control. Chapter five concerns horticultural crops and, in addition to some preliminary operations, describes a number of combined optional treatments such as irradiation, refrigeration, pickling, and packaging, transport and quality control.

Fruits and vegetables represent an important and in many cases an under-appreciated resource which could benefit from better utilisation and exploitation in the rural communities. This manual has therefore been designed as a useful reference book for food producers, traders and processors. Other users include extension agents and rural development practitioners active in the processing and preservation aspects of the food chain.

Geoffrey C. Mrema
Director
Agricultural Support Systems Division
Food and Agriculture Organization

ACKNOWLEDGEMENTS

The authors appreciate and acknowledge Dr. Danilo J. Mejía, Officer of Agricultural and Food Engineering Technologies Service (AGST, FAO, Rome), not only for the Table of Contents he proposed to the authors, but also for his invaluable help throughout the generation and editing of this manual. The authors also want to acknowledge the CYTED (Ibero-American Program to Promote Science and Technology), Subprogram XI for many years of support in developing and promoting the combined methods technology for fruits and vegetables, as well as other commodities.

HANDLING AND PRESERVATION OF FRUITS AND VEGETABLES BY COMBINED METHODS FOR RURAL AREAS

INTRODUCTION

This manual presents information related to the processing of fruits and vegetables by combined methods. It is intended to serve as a guide to farmers and processors of fruits and vegetables in rural and village areas. Information concerning the trade and production of fruits and vegetables in different countries is provided, as well as information on the processing of fruit and vegetable products. The combination of factors such as water activity (a_W), pH, redox potential, temperature, and incorporation of additives in preserving fruits and vegetables is important, and all play a crucial role in improving the shelf life of fresh and processed commodities.

The increasing popularity of minimally processed fruits and vegetables has resulted in greater health benefits. Furthermore, the ongoing trend has been to eat out and to consume ready-to-eat foods (Alzamora et al., 2000). With this increasing demand for ready-to-eat, fresh, minimally processed foods, including processed fruits and vegetables preserved by relatively mild techniques, new ecology routes for microbial growth have emerged. In order to minimize the loss of quality and to control microbial growth, and thus ensure product safety and convenience, a hurdle approach appears to be the best method (Alzamora et al., 2000). According to Alzamora et al. (2000), hurdle technology can be applied several ways in the design of preservation systems for minimally processed foods at various stages of the food chain:

- As a "backup" measure for existing minimally processed products with short shelf life, to diminish microbial pathogenic risk and/or increase shelf life (i.e., use of natural antimicrobials or other stress factors, in addition to refrigeration).
- As an important tool for improving the quality of long shelf life products without diminishing their microbial stability/safety (i.e., use of heat coadjuvants to reduce the severity of thermal treatments).
- As a synergist. According to Leistner (1994), in food preserved by hurdle technology, the possibility exists that different hurdles in a food will not just have an additive effect on stability, but could act synergistically. A synergist effect could work if the hurdle in a food hits different targets (e.g., cell membrane, DNA, enzyme systems, pH, a_W, Eh) within the microbial cell, and thus disturbs the homeostasis of the microorganisms present in several aspects. Therefore, employing different hurdles in the preservation of a particular food should be an advantage, because microbial stability could be achieved with a combination of gentle hurdles. In practical terms, this could mean that it is more effective to use different preservatives in small amounts in a food than only one preservative in large amounts, because different preservatives might hit different targets within the bacterial cell, and thus act synergistically (Leistner, 1994).

During the last decade, minimally processed high moisture fruit products (HMFP), which are ambient stable (with $a_W > 0.93$), have been developed in seven Latin American countries, under the leadership of Argentina, Mexico, and Venezuela. This novel technology was successfully applied to peach halves, pineapple slices, mango slices and purée, papaya slices, chicozapote slices, banana purée, plum, passion fruit, tamarind, whole figs, strawberries, and pomalaca (Alzamora et al., 1995). The methodology employed was based on combinations of mild heat treatments, such as blanching for 1-3 minutes with saturated steam, slightly reducing the a_W (0.98-0.93) by addition of glucose or sucrose, lowering the pH (4.1-3.0) by addition of citric or phosphoric acid, and adding antimicrobials (1000 ppm of potassium sorbate or sodium benzoate, as well as 150 ppm of sodium sulphite or sodium bisulphite) to the product syrup. During storage of HMFP, the sorbate and sulphite levels decreased, as well as a_w levels, due to hydrolysis of glucose (Alzamora et al., 1995).

The work presented in this manual demonstrates at which stage of maturity a fruit or vegetable should be harvested, and packaged, for optimum storability, marketable life, quality, and all aspects related to final use of fresh and processed products. Some useful examples, figures, and tables concerning the preservation of fruits and vegetables by combined methods are demonstrated

This book also summarizes the basic principles of harvest and post-harvest handling and storage of fresh fruits and vegetables.

CHAPTER 1
FRUITS AND VEGETABLES: AN OVERVIEW ON SOCIO-ECONOMICAL AND TECHNICAL ISSUES

1.1. Trade and global trends: fruits and vegetables

Recently, the Food Agricultural Organization of the United Nations (FAO) predicted that the world population would top eight billion by the year 2030. Therefore, the demand for food would increase dramatically. As stated in the FAO report, "Agriculture: Towards 2015/30", remarkable progress has been made over the last three decades towards feeding the world. While global population has increased over 70 percent, per capita food consumption has been almost 20 percent higher. In developing countries, despite a doubling of population, the proportion of those living in chronic states of under nourishment was cut in half, falling to 18 percent in 1995/97. According to the report, crop output is projected to be 70 percent higher in 2030 than current output. Fruits and vegetables will play an important role in providing essential vitamins, minerals, and dietary fibre to the world, feeding populations in both developed and developing countries.

In developed countries, the U.S. continues to dominate the international trade of fruits and vegetables, and is ranked number one as both importer and exporter, accounting for approximately 18 percent of the $40 billion (USD) in fresh produce world trade. As a group, the European Union (EU) constitutes the largest player, with 15 additional export and import commodities contributing about 20 percent to total fresh fruit and vegetable trade. Within Europe, Germany is the principal exporter; Spain is the principal supplier; and the Netherlands plays an important role in the physical distribution process. In the Southern Hemisphere, Chile, South Africa, and New Zealand have become major suppliers in the international trade of fresh fruit commodities, although they remain insignificant in vegetable trade.

FAO estimated that the world production of fruits and vegetables over a three-year period (1993-1995) was 489 million tons for vegetables and 448 million tons for fruits. This trend increased as expected, reaching a global production of 508 million tons for vegetables and 469 tons for fruits in 1996. This trend in production is expected to increase at a rate of 3.2 percent per year for vegetables and 1.6 percent per year for fruits. However, this trend is not uniform worldwide, especially in developing countries where the lack of adequate infrastructure and technology constitutes the major drawback to competing with industrialized countries. Nevertheless, developing countries will continue to be the leaders in providing fresh exotic fruits and vegetables to developed countries. Most developing countries have experienced a high increase in fruit and vegetable production, as in the case of Asia (China) and South America (Brazil, Chile). Asia is the leading producer of vegetables with a 61 percent total volume output and a yearly growth of 51 percent. However, the U.S. continues to lead in the export of fresh fruits and vegetables worldwide with orange, grapes, and tomatoes. Brazil dominates the international trade of frozen orange juice concentrate, while Chile has become the major fresh fruit exporter with a production volume of 45 percent. Despite the large growth in exports in the 1990s, the U.S. remains a net importer of horticultural products. As U.S. consumers have become more willing to try new fruit and vegetable varieties, the imported share of the domestic market has increased. According to a USDA report, the total value of horticultural products imported into the U.S. has grown by more than 50 percent since 1990. If long-term projections hold for the next decade, the U.S. could achieve a trade balance surplus in horticultural products, due mainly to a global increase in the market. While the import value of horticultural products is projected to grow at a steady rate of 4 percent per year, between 1998 and 2007, the USDA's baseline projection period for exports are projected to grow by 5 to 7 percent per year.

The top six fruit producers, in declining order of importance, are China, India, Brazil, USA, Italy, and Mexico. China, India, and Brazil account for almost 30 percent of the world's fruit supply, but since most of this production is destined for domestic consumption its impact on world trade is minimal.

1.2. Traditional consumption

Fruit and vegetable consumption per capita showed an increase of 0.38 percent for fresh fruits and 0.92 percent for vegetables per capita from 1986 to 1995. The highest consumption of fresh fruits was registered in China (6.4%), as the apparent per capita consumption of vegetables in China went from 68.7 kg per capita in 1986 to 146 kg in 1995 (53.8% growth rate), while African and Near East Asian countries showed a decrease in fresh fruit consumption. The lowest consumption of vegetables per capita was registered in Sub-Saharan Africa (29 kg of vegetables consumed both in 1986 and 1995). According to trade sources, Chinese customers purchased most of their fresh fruit at street retail shops and market places where imported fresh fruits are available and U.S. and European brand names have received recognition. Products such as Red Delicious apples, Sunkist oranges, and Red Globe table grapes are especially popular. Sunkist is one of the few brands of oranges consumers recognize. The trend toward fresh vegetable consumption in developing countries is one indication of the population's standard of living, but generally, fresh vegetables lose their market share to processed products. Many vegetables can be processed into canned products that cater to local tastes, (e.g., cucumbers and peppers). Easy to carry and convenient to serve, they can be stored for a long time, reducing losses incurred from the seasonal supply of surplus vegetables marketed yearly at the same time. Urban population is exploding in developing countries, having risen from 35 percent of the total population in 1990, and projected to rise 54 percent in 2020. With increasing urban populations, more free markets and wholesale markets will be required to increase the supply of fresh fruits and vegetables. For example, the growth of consumption in the U.S. has been stimulated partly by increasing demand for tropical and exotic fruits and vegetables (mainly imported).

1.3 Economic and social impact

Ongoing consumer demand for new fruits and vegetables in developed countries has contributed to an increase in trade volume of fresh produce in developing countries. This, in turn, has promoted the growth of small farms and the addition of new products, creating more rural and urban jobs and reduced the disparities in income levels among farms of different sizes. As countries become wealthier, their demand for high-valued commodities increases. The effect of income growth on consumption is more pronounced in developing countries, compared to developed countries, they are expected to spend larger shares of extra income on food items like meat and fruit and vegetable products. The implementation of international trade agreements, such as NAFTA (U.S., Mexico, Canada) and MERCOSUR (Argentina, Brazil, Paraguay and Uruguay), has significantly impacted the economy of the signatory countries by increasing the trade volumes and trade flows, particularly through general areas such as market access, tarification, limits on export subsidies, cuts in domestic supports, phyto-sanitary measures, and safeguard clauses.

1.4 Commercial constraints

According to the USDA economic report, the commercial constraints on fruits and vegetables include:

Trade barriers: Natural and artificial barriers. Natural trade barriers include high transportation costs to distant markets, and artificial barriers include legal measures such as protectionist policies. Liberalization of trade through international agreements has been instrumental in relaxing many legal trade barriers by reducing tariffs and by harmonizing the technical barriers to trade.

Scientific phyto-sanitary requirements: Importing countries set the standards that potential trade partners must meet in order to protect human health or prevent the spread of pests and diseases. For instance, Japanese imports of U.S. apples are limited to Red and Golden Delicious apples from Washington and Oregon. The Japanese, who are mainly concerned with the spread of fire blight, impose rigorous and costly import requirements on the U.S. apple shippers. The apples must be subjected to a cold treatment and fumigation with methyl bromide before shipment to Japan, and three inspections of U.S. apple orchards during the production stage. Infestation by fruit flies (*Tepbritidae: Diptera*), common in the tropics, is a major constraint to the production and export of tropical fruits.

Technological innovations: Countries can increase their competitiveness and world market shares by providing higher quality products and promoting lower prices through technological innovations.

Trade liberalization, negotiated through the Uruguay Round Agreement (URA) (of the GATT and implemented under WTO), as well as through regional agreements, such as NAFTA and MERCOSUR, has expanded market access and provided strengthened mechanisms for combating non-tariff trade barriers such as scientifically unfounded phyto-sanitary restrictions. Future prospects of fruits and vegetables exported from developing countries will largely depend on the growth of import demand, mostly in the developed countries. Developed countries are expected to diversify their consumption of fruits and vegetables. This will increase the concern about health and nutrition; the consumer's familiarity with more fruits and vegetables because of wider availability, increased travel, and improved communications will lead to an increase in the ratio of imports to domestic products (Segre, 1998).

1.5 Post-harvest losses and resource under-utilization in developing countries

Postharvest losses of fruits and vegetables are difficult to predict; the major agents producing deterioration are those attributed to physiological damage and combinations of several organisms. Flores (2000) described postharvest losses due to various causes as follows:

1.5.1 Food losses after harvesting These include losses from technological origin such as deterioration by biological or microbiological agents, and mechanical damage.

Losses due to technological origin include: unfavourable climate, cultural practices, poor storage conditions, and inadequate handling during transportation, all of which can lead to accelerated product decay (e.g., tubers re-sprouting from bulbs and weight loss from product dehydration).

Physiological deterioration of fruits and vegetables refers to the aging of products during storage due to natural reactions. Deterioration caused by biochemical or chemical agents refers to reactions, of which intermediate and final products are undesirable. These can result in significant loss of nutritional value (i.e., rancidity and agrochemical contamination) and in many cases the whole fruit or vegetable is lost.

Deterioration by biological or microbiological agents refers to losses caused by insects, bacteria, moulds, yeasts, viruses, rodents, and other animals. When fruits and vegetables are gathered into boxes, crates, baskets, or trucks after harvesting, they may be subject to cross-contamination by spoilage microorganisms from other fruits and vegetables and from containers.

Most of the microorganisms present in fresh vegetables are saprophytes, such as *coryniforms,* lactic acid bacteria, spore-formers, *coliforms, micrococci*, and pseudomonas, derived from the soil, air, and water. *Pseudomonas* and the group of *Klebsiella-Enterobacter-Serratia* from the enterobacteriaceae are the most frequent. Fungi, including *Aureobasidium, Fusarium*, and *Alternaria*, are often present but in relatively lower numbers than bacteria. Due to the acidity of raw fruits, the primary spoilage organisms are fungi, predominantly moulds and yeasts, such as *Sacharomyces cerevisiae*, *Aspergillus niger, Penicillum spp., Byssochlamys fulva, B. nivea*, *Clostridium pasteurianum*, *Coletotrichum gloesporoides*, *Clostridium perfringes*, and *Lactobacillus spp*. Psychrotrophic bacteria are able to grow in vegetable products; some of them are *Erwina carotovora*, *Pseudomonas fluorescens, P. auriginosa, P. luteola, Bacillus species, Cytophaga jhonsonae, Xantomonas campestri*, and *Vibrio fluvialis* (Alzamora et al., 2000).

The existence of pathogenic bacteria in fresh fruit and vegetable products has been reported by Alzamora et al. (2000), which include *Listeria monocytogenes, Aeromonas hydrophila,* and *Escherichia coli* O157: H7. These bacteria are found in both fresh and minimally processed fruit and vegetable products. *Listeria monocytogenes* is able to survive and grow at refrigeration temperatures on many raw and processed vegetables, such as ready-to-eat fresh salad vegetables, including cabbage, celery, raisins, fennel, watercress, leek salad, asparagus, broccoli, cauliflower, lettuce, lettuce juice, minimally-processed lettuce, butterhead lettuce salad, broad-leaved and curly-leaved endive, fresh peeled hamlin oranges, and vacuum-packaged potatoes (Alzamora et al., 2000). *Aeromonas hydrophila* is a characteristic concern in vegetables; it is a psychrotrophic and facultative anaerobe. *Aeromonas* strains are susceptible to disinfectants, including chlorine, although recovery of *Aeromonas* from chlorinated water has been reported. Challenging studies inoculating *A. hydrophila* in minimally processed fruit salads showed that *A. hydrophila* was able to grow at 5°C during the first 6 days, however, the pathogen decreased after 8 days of storage. (Alzamora et al., 2000). *E. coli O157H:7* has emerged as a highly significant food borne pathogen. The principal reservoir of E. *coli O157H:7* is believed to be the bovine gastrointestinal tract. Thus, contamination of associated food products with faeces is a significant risk factor, particularly if untreated contaminated water is consumed directly or used to wash uncooked foods.

Mechanical damage is caused by inappropriate methods used during harvesting, packaging, and inadequate transporting, which can lead to tissue wounds, abrasion, breakage, squeezing, and escape of fruits or vegetables. Mechanical damage may increase susceptibility to decay and growth of microorganisms. Some operations, such as washing, can reduce the microbial load; however, they may also help to distribute spoilage microorganisms and moisten surfaces enough to permit growth of microorganisms during holding periods (Alzamora et al., 2000). All methods of harvesting cause bruising and damage to the cellular and tissue structure, in which enzyme activity is greatly enhanced as cellular components are dislocated (Holdsworth, 1983).

Besides the above issues, most post-harvest losses in developing countries occur during transport, handling, storage, and processing. Rough handling during preparation for market will increase bruising and mechanical damage, and limits the benefits of cooling.

By-products from fruit and vegetable processing are not wholly utilized in developing countries due to lack of machinery and infrastructure to process waste. The easiest way to dispose of by-products is to dump the waste or use it directly as animal feed. Waste materials such as leaves and tissues could be used in animal feed formulations and plant fertilizers.

In general, it is estimated that between 49 to 80% goes to consumers in the production of a particular commodity, and the difference is lost during the varied steps that comprise the harvest-consumption system.

1.5.2 Food losses due to social and economic reasons

Policies: This involves political conditions under which a technological solution is inappropriate or difficult to put in to practice, for example, lack of a clear policy capable of facilitating and encouraging utilization and administration of human, economic, technical, and scientific resources to prevent the deterioration of commodities.

Resources: This is related to human, economic, and technical resources for developing programs aimed at prevention and reduction of post-harvest food losses.

Education: This includes unknown knowledge of technical and scientific technologies associated with preservation, processing, packaging, transporting, and distribution of food products.

Services: This refers to inefficient commercialization systems, and absent or inefficient government agencies in the production and marketing of commodities, as well as a lack of credit policies that address the needs of the country and participants.

Transportation: This is a serious problem faced by fruit growers in developing countries, where vehicles used in transporting bulk raw fruits to markets are not equipped with good refrigeration systems. Raw fruits exposed to high temperatures during transportation soften in tissue and bruise easily, causing rapid microbial deterioration.

1.6 Pre-processing to add value

Rapid cooling of produce following harvest is essential for crops intended for transport in refrigerated ships, land vehicles, and containers not designed to handle the full load of field heat but capable of maintaining precooled produce at a selected carriage temperature. The selected method of cooling will depend greatly on the anticipated storage life of the commodity. Rapidly respiring commodities with short post-harvest life should be cooled immediately after harvest. Therefore, added value is achieved in precooling the produce immediately after harvest, which will restrict deterioration and maintain the produce in a condition acceptable to the consumer.

Blanching of fruits as a pre-treatment method may also be applied before freezing and juicing, or in some cases, before dehydration (Arthey and Ashurst, 1996). The fruit may be blanched either by exposure to near boiling water, steam, or hot air for 1 to 10 minutes. Blanching inactivates those enzyme systems that degrade flavour and colour and cause vitamin loss during subsequent processing and storage (Arthey and Ashurst, 1996).

1.7 Pre-processing to avoid losses

Pre-processing of fruits and vegetables includes: blanching to inactivate enzymes and microorganisms, curing of root and tubers to extend shelf life, pre-treatment of produce with cold or high temperatures, and chemical preservatives to control pests after harvest. Storage of produce under controlled temperature and relative humidity conditions will extend its perishability and reduce decay. Packaging of produce in appropriate material enhances colour appearance and marketability.

1.8 Alternative processing methods for fruits and vegetables in rural areas

A variety of alternative methods to preserve fruits and vegetables can be used in rural areas, such as fermentation, sun drying, osmotic dehydration, and refrigeration.

Fruits and vegetables can be pre-processed via scalding (blanching) to eliminate enzymes and microorganisms. Fermentation of fruits and vegetables is a preservation method used in rural areas, and due to the simplicity of the process, there is no need for sophisticated equipment; pickled produce, sauerkraut, and wine are examples of this process. A general schematic diagram of the different alternative processes for fruits and vegetables is presented in Figure 1.1, and described as follows:

Cleaning and washing are often the only preservation treatments applied to minimally processed raw fruits and vegetables (MPRFV). As the first step in processing, cleaning is a form of separation concerned with removal of foreign materials like twigs, stalks, dirt, sand, soil, insects, pesticides, and fertilizer residues from fruits and vegetables, as well as from containers and equipment. The cleaning process also involves separation of light from heavy materials via gravity, flotation, picking, screening, dewatering, and others (Wiley, 1996). Washing is usually done with chlorinated water (i.e., 200 ppm allowed in the USA). The MPRFV product is immersed in a bath in which bubbling is maintained by a jet of air. This turbulence permits one to eliminate practically all traces of air and foreign matter without bruising the product.

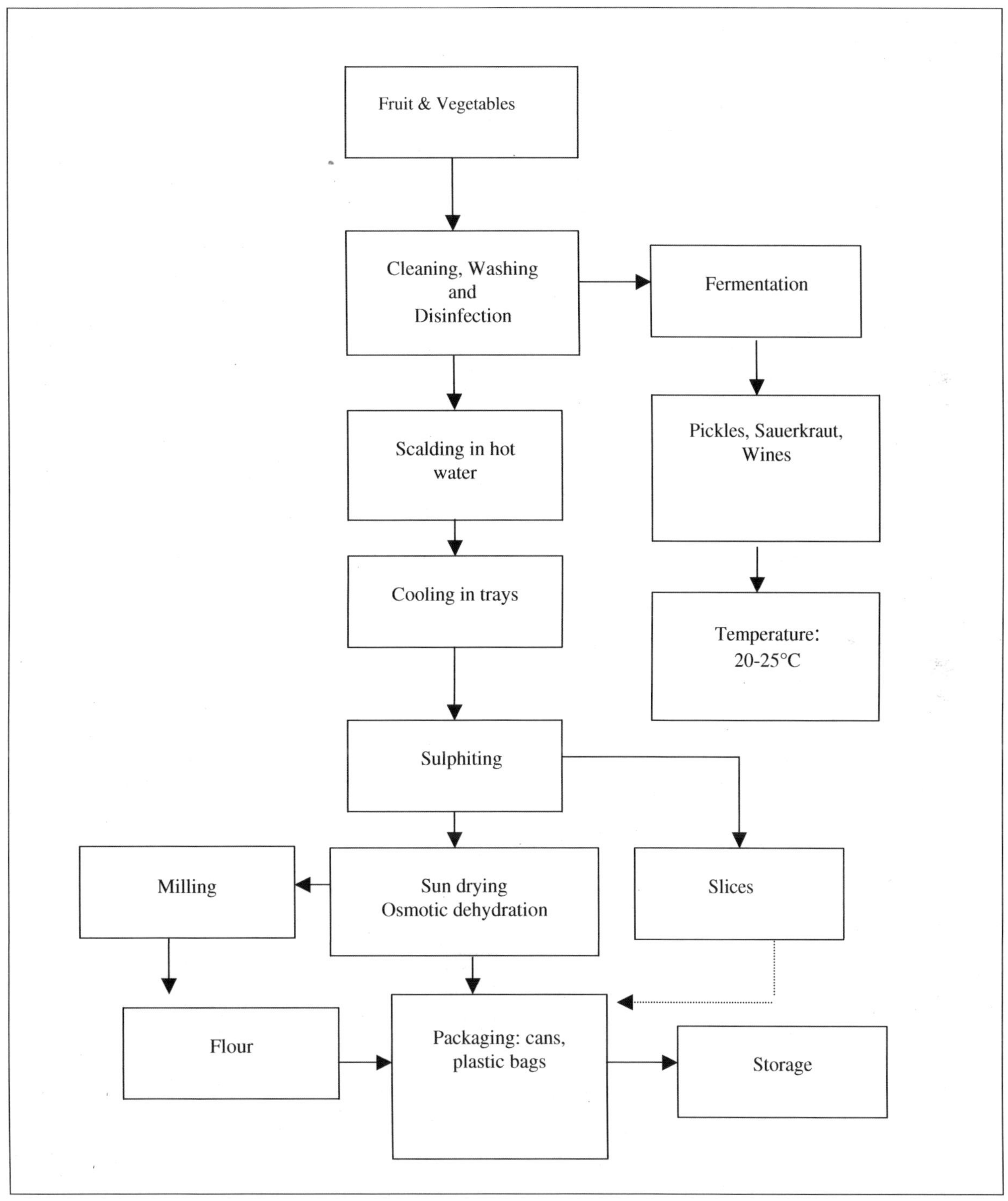

Figure 1.1 Processing of fruits and vegetables in rural areas.

Water must be of optimal quality for washing MPRFV products, otherwise cross contamination may occur. According to Wiley R.C. (1997), three parameters are controlled in washing MPRFV fruits and vegetables:

1. Quantity of water used: 5-10 L/kg of product
2. Temperature of water: 4°C to cool the product
3. Concentration of active chlorine: 100 mg/L

Two examples of specially designed equipment used to wash fruits and vegetables include: 1) rotary drums used for cleaning apples, pears, peaches, potatoes, turnips, beets; high pressure water is sprayed over the product, which never comes in contact with dirty water, and 2) wire cylinder leafy vegetable washers, in which medium pressure sprays of fresh water are used for washing spinach, lettuce, parsley, and leeks.

In rural areas, fresh produce could be poured into plastic containers filled with tap water to remove the dirt from fruits and vegetables. The dirty water could be drained from the containers and refilled with chlorinated water for rewashing and disinfection of the fruit or vegetable. If electricity is available, fresh produce could be refrigerated until processed or distributed to retailers and markets.

1.8.1 Scalding or blanching in hot water:

Fruits, fresh vegetables and root vegetable pieces are immersed in a bath containing hot water (or boiling water) for 1-10 minutes at 91-99°C, to reduce microbial levels, and partially reduce peroxidase and polyphenoloxydase (PPO) activity. The heating time will depend on the type of vegetable product processed Boiling water has been used to provide thermal inactivation of *L. monocytogenes* on celery leaves (Wiley, 1997).

1.8.2 Cooling in trays:

This operation is carried out in perforated metal trays through which cool air is passed in order to cool the product prior to packaging in sterile plastic bags, unless another process is to follow.

1.8.3 Sulphiting

During this operation, the fruit or vegetable pieces (or slices) are immersed in a solution of sodium bisulphite (200 ppm) to prevent undesirable changes in colour and any additional microbial and enzyme activity, and to retain a residual concentration of 100 ppm in the final product.

1.8.4 Sun drying and Osmotic dehydration

In rural areas, dehydration is probably the most effective method to preserve fruits and vegetables. Fruit slices or vegetable pieces are spread over stainless metal trays or screens spaced 2-3 cm apart and sun dried. The dried fruit and vegetable products are then packaged in plastic bags, glass bottles, or cans, as with fruit slices (i.e., mango, papaya, peach, etc.) or milled flour (i.e., green plantain flour produced in rural areas of developing countries).

In osmotic dehydration and crystallization, the fruit is preserved by heating the product in sugar syrup, followed by washing and drying to reduce the sugar concentration at the fruit surface. Fruits are dried by direct or indirect sun drying, depending on the quality of the product obtained. The advantage of this method is the prevention of discoloration and browning of fruit produced by enzymatic reactions. Thus, the high concentration of sugar in the fruit produces a dehydrated product with good colouring, without the need of chemical preservatives such as sulphur dioxide.

1.8.5 Fermentation

This is another useful preservation process for fruit and vegetable products. For vegetables, the product is immersed into a sodium chloride solution, as in the case of cucumbers, green tomatoes, cauliflower, onions, and cabbage (sauerkraut). Composition of the salt (sodium chloride) is maintained at about 12% by weight so that active organisms during fermentation, such as Lactic acid bacteria, and the *Aerobacter* group, produce sufficient acid to prevent any food poisoning organisms from germinating (Holdsworth, 1983). Fruits, on the other hand, can be preserved by fermenting the fruit pulp into wine, by preparing a solution of sugar and water and then inoculating it with a strain of *Saccharomyces cerevisiae*. This process is very simple and will be discussed in greater detail later in this chapter.

1.8.5.1 Pickles, sauerkraut and wine making

Slightly underripe cucumbers are selected and cleaned thoroughly with water, then size-graded prior to brining. For a large production of pickles, the fermentation process is carried out in circular wooden vats 2.5-4.5 m in diameter and 1.8-2.5 m deep. A small batch of pickles can be produced using appropriate plastic containers capable of holding 4-5 kg of cucumbers. After the cucumbers are put into the vats, a salt solution (approximately 10% by weight) is added. This concentration is maintained by adding further salt as needed by recirculating the solution to eliminate concentration gradients. Sugar is added if the cucumbers are low in sugar content to sustain the fermentation process (Holdsworth, 1983). The fermentation process will end after 4-6 weeks, and the salt concentration will rise to 15%. Under these conditions, pickles will keep almost indefinitely. Care must be taken to ensure that the yeast scum on top of the vat does not destroy the lactic acid. This can be done by adding a layer of liquid paraffin on the surface of the pickling solution. After the fermentation process has ended, the pickles are soaked in hot water to remove excess salt, then size-graded and packed into glass jars with acetic acid in the form of vinegar. A flow diagram for this process is illustrated in Figure 1.2 (see page 12).

1.8.5.2 Sauerkraut

Selected heads of cabbage are core-shredded and soaked in tap water with 2.5% (by weight) salt concentration and allowed to ferment. During the initial stages of fermentation, there is a rapid evolution of gas caused by *Leuconostoc mesenteroides;* this process imparts much of the pleasant flavour to the product. The next stage involves *Lactobacillus cucumeris* fermentation, resulting in an increase of lactic acid; and finally after approximately 5 days at 20-24°C, the third stage, involving a further group of lactic acid bacteria such as *Leuconostoc pentoaceticus*, which yields more lactic acid combined with acetic acid, ethyl alcohol, carbon dioxide, and mannitol. The fermentation process ends when the lactic acid production is approximately 1-2%. This can be tested by titration of the acid with a 0.1 N sodium hydroxide (NaOH) solution, using phenolphthalein (0.1% w/v) as colour indicator (i.e., 2-5 drops are added to the acid solution; colour will change from clear to pink and persists for 30 seconds). After the fermentation process, either the tank is sealed to exclude air or the product is then packed into glass jars or canned. It is then ready for consumption (Figure 1.3 see page 13). Further details regarding sauerkraut production are given in Chapter 5.

1.8.5.3 Wine making

Selected ripened fruits are transported to the farm where they are sorted, washed and macerated or chopped prior to pressing. In rural areas, juice is extracted from the fruit by squeezing (oranges, grapes, etc.) or pulped (mangoes, maracuyá, guava, etc.). The soluble solid content of the pulp is measured with a refractometer in °Brix. Soluble solids should be 25%, but if lower, it can be adjusted with sugar.

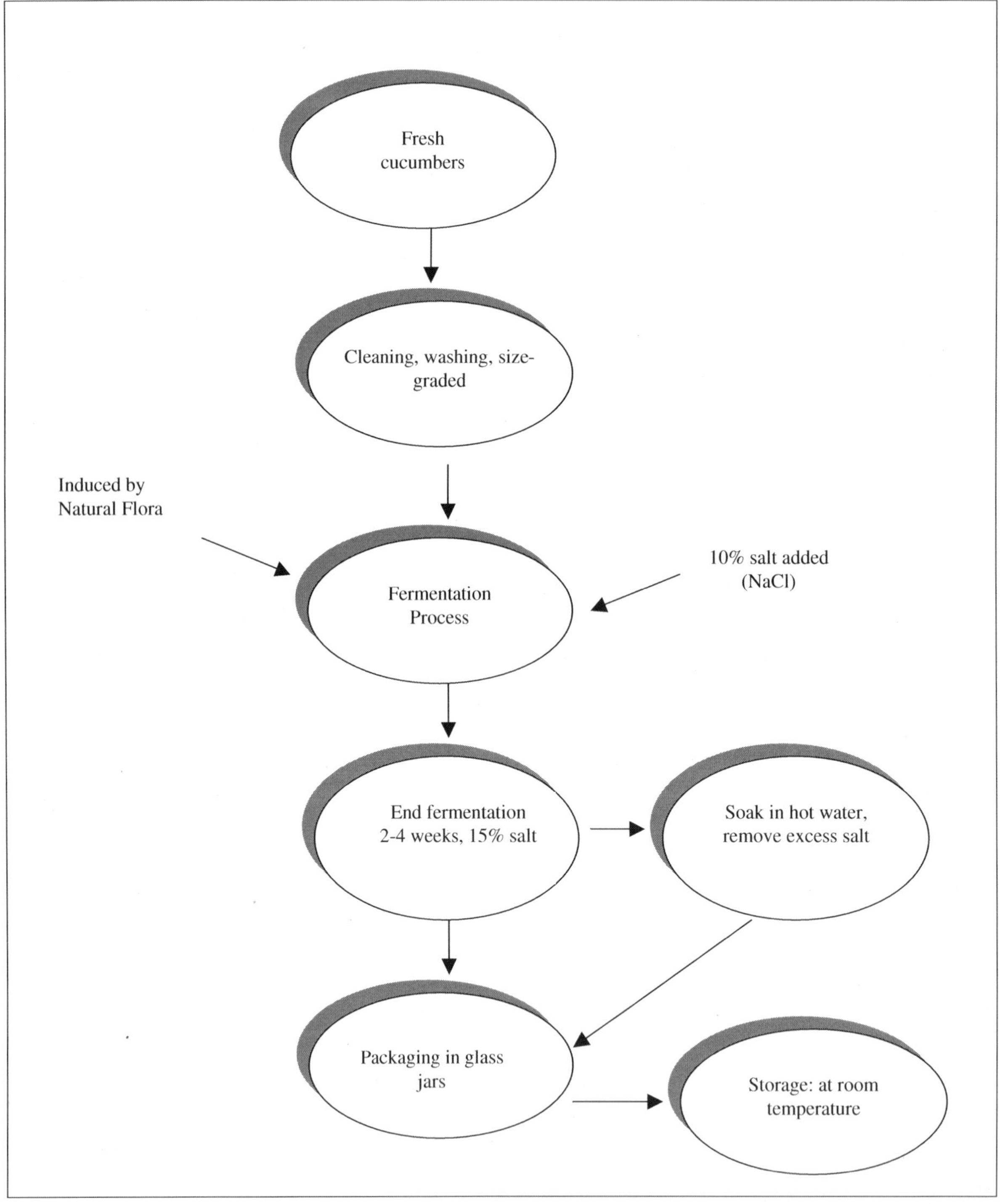

Figure 1.2. Flow diagram for pickle production.

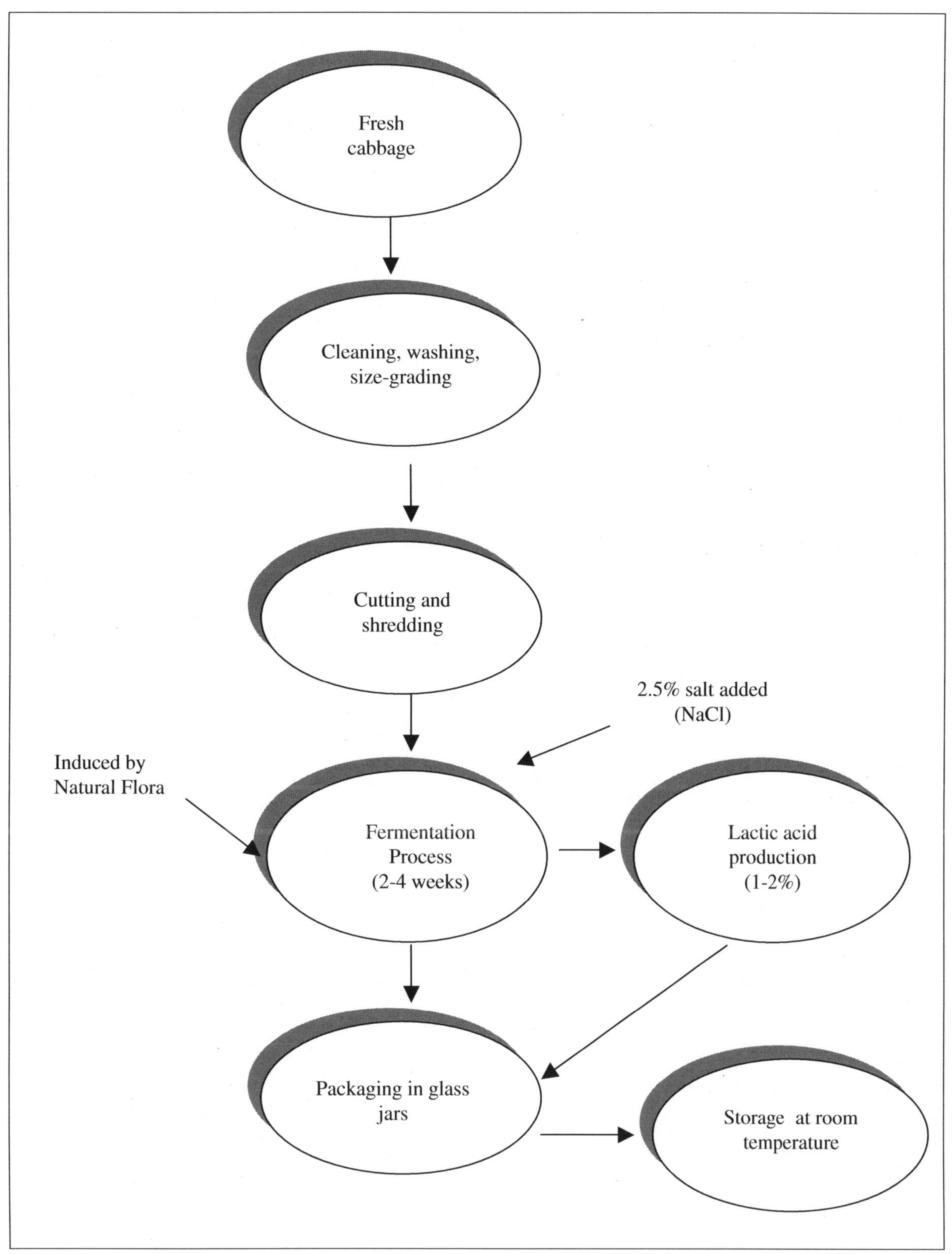

Figure 1.3. Flow diagram for sauerkraut production.

Clarification:

Clarification of wines prior to bottling involves treatment with gelatine, albumin, isinglass, bentonite, potassium ferrocyanide or salts (the last two treatments are intended to reduce the level of soluble iron complexes, which would otherwise cause a darkening of the wine, but with fruit wine these are frequently inadequate (Arthey and Ashurst, 1996). Alternative clarification procedures include chilling the wine prior to, or after, refining, and using microfiltration systems. A simple way to clarify wine is to add white gelatine (1 g per L of wine) to the fermented fruit solution, which is then allowed to stand in the refrigerator for 1 week, after which all of the suspended solids are precipitated and a clear transparent wine can be decanted from the top of the container. Following clarification, the wine will normally be flash pasteurized, hot-filled into bottles, or treated to give a residual SO_2 content (100 ppm).

The next stage is to add sodium bisulphite to the fruit juice (200 ppm), allowing it to stand for 2-3 hours. During this process, the unwanted yeast flora present in the fruit pulp is eliminated and the added inoculum can act freely in the fruit juice to produce the desired flavour or bouquet characteristic of fruit wine. Next, the yeast is added to the juice (1 g per kg of fruit juice, usually strains of *Saccharomyces cerevisiae* or bread making yeast). The fermentation process should be carried out anaerobically, that is in the absence of oxygen, to prevent development of other non-wine making bacteria, such as *Acetobacter spp,* which produces undesirable taste and flavour. The fermentation ends after 3 to 4 weeks at 22-25°C.

The final stage of processing involves the blending, sweetening and flavouring (if required), and stabilization of the wines. The blending process is done both to ensure consistency of product character and to reduce the strong aroma and flavour of certain wines. Although there is some preference for single wines, many are blended, especially with apple wine, which is relatively low in flavour. Wines can be sweetened using sugar or fruit juice, the latter also serving to increase the natural fruit content. In some cases, it is necessary to adjust the acidity of wine by adding an approved food-grade acid, such as citric or tartaric acid. In many rural areas, where these chemicals are not available, lemon juice can be used instead. Flow diagrams for this process are shown in Figure 1.4.

For wine making in rural areas, the fermentation process is usually carried out in a large bottle (18-20 L), in which the ingredients are mixed with water. In order to keep the fermentation process under anaerobic conditions, a water-filled air-lock is fitted into a hollow cork or rubber stopper inside the mouth of the bottle. This can be made simply from a piece of plastic tubing and a bottle (Figure 1.5 see page 17).

1.8.6 Storage

Because sun dried and fermented fruit and vegetable products are stable, they can be stored at ambient temperatures or at low refrigeration temperatures, extending the shelf life for several months (6-12 months and beyond). Wine is stored in glass bottles and maintained at room temperature or it can be stored under refrigeration. Other fermented products such as sauerkraut and pickles are usually stored at room temperature.

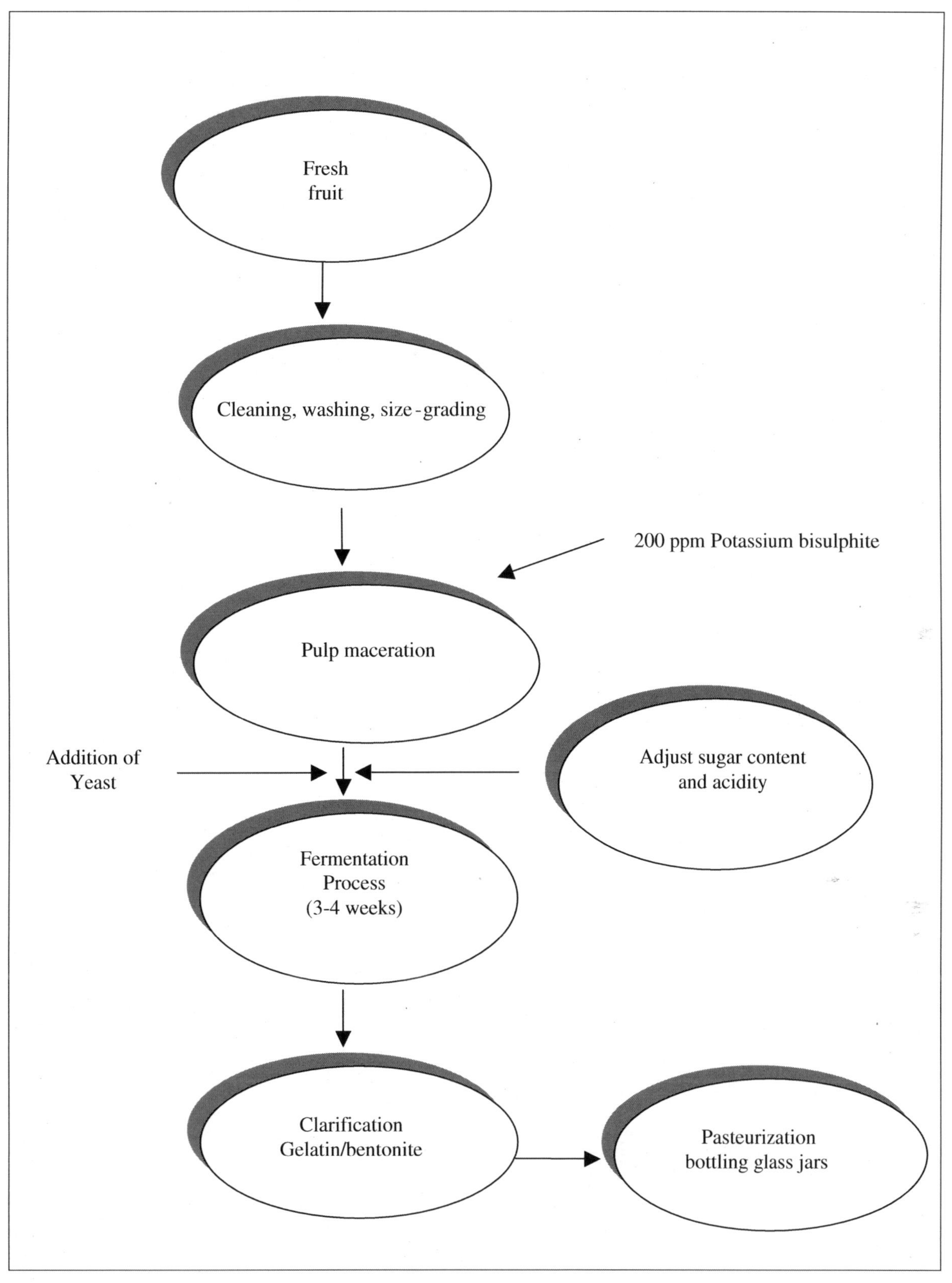

Figure 1.4. Flow diagram for fruit wine production.

1.8.7 Sample calculation for adjusting fruit soluble solids and acid contents

Ten kg of fruit pulp contains approximately 5% soluble solids (i.e., 5 kg sugar/100 kg pulp) and 0.2% citric acid. We want to adjust the soluble solids to 25% and 0.7% citric acid.

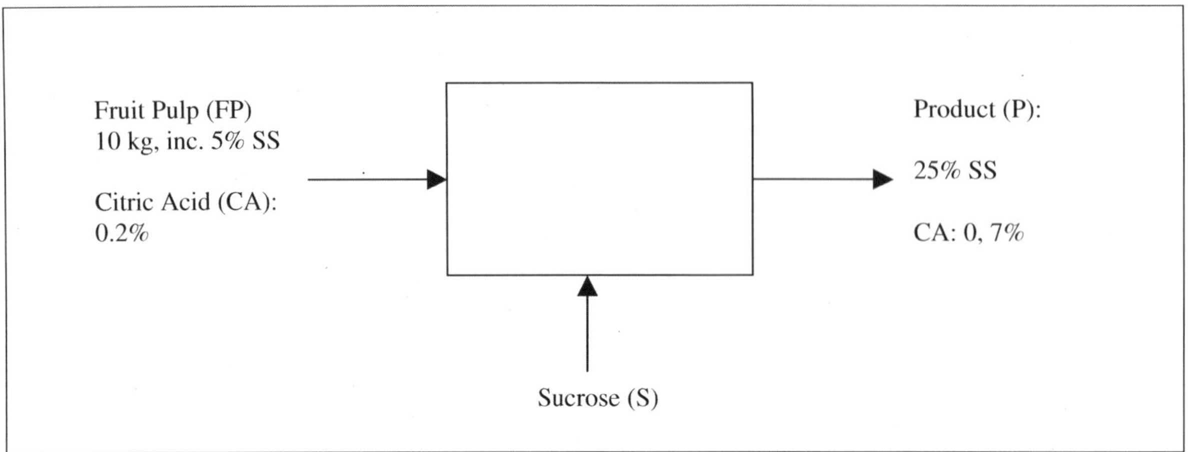

Solution:

Total Material Balance:

$FP + S = P$ (1)

Balance of soluble solids:

$10\ (0.05) + S = 0.25P$ (2)

From (1): $S = P - 10$

Substituting S into (2), we get:

$10\ (0.05) + P - 10 = 0.25P$

$0.75P = 9.5$

Solving for P,

$P = 9.5/0.75 = 12.67$ kg

Therefore, the amount of sucrose needed to adjust the soluble solids content in fruit pulp from 5 % to 25% equals:

$S = P - 10 = 12.67 - 10 =$ **2.67 kg of Sucrose**

Citric acid adjustment:

Acid Balance:

$10(0.002) + CA = 12.67(0.007)$

$CA = 0.08869 - 0.02 =$ **0.0687 kg of Citric Acid**

Final product will have the following composition:

10 kg fruit pulp + 2.67 kg sucrose (25% solids content)

0.0687 kg Citric Acid (7% citric acid content)

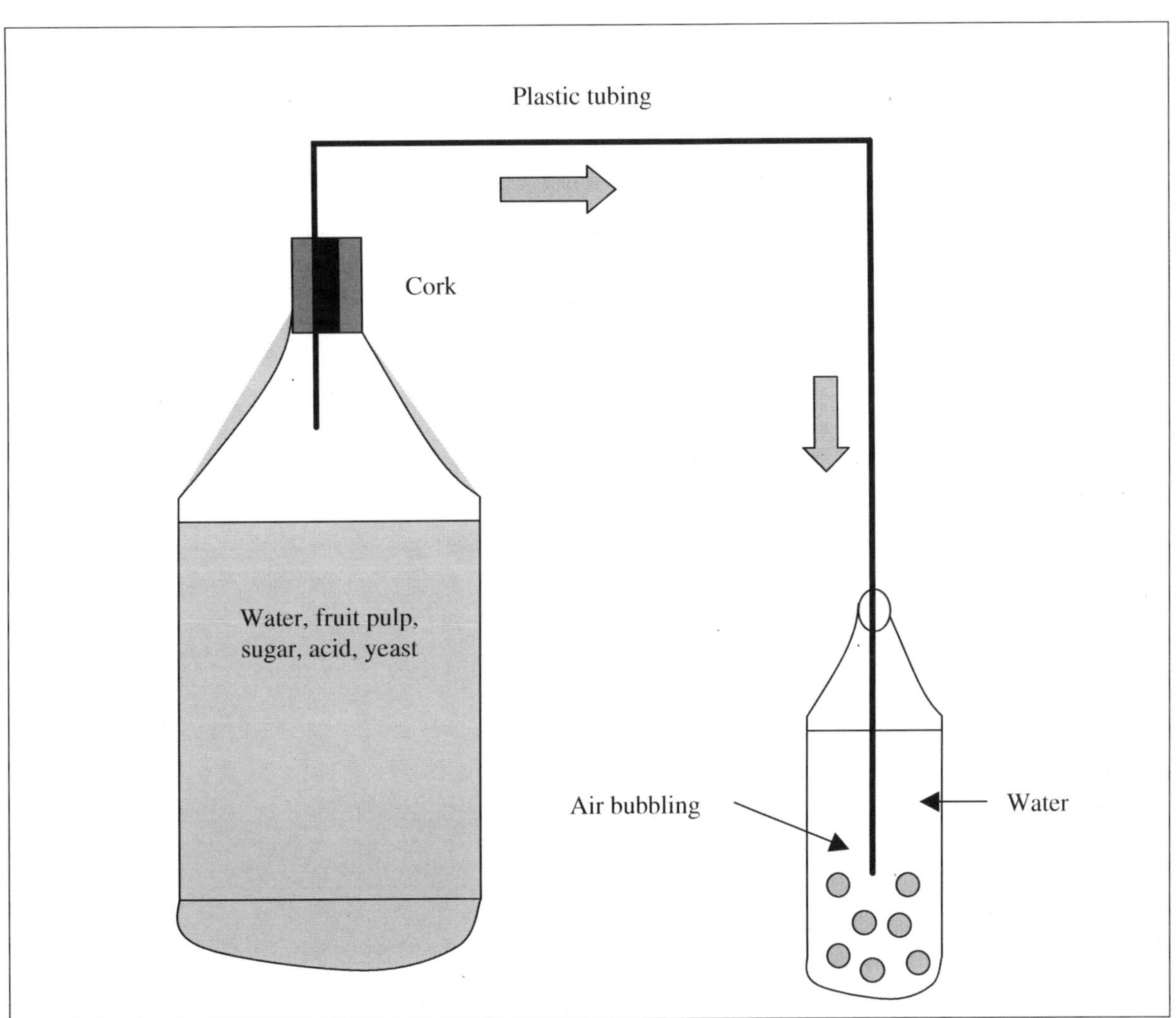

Figure 1.5. Schematic diagram for wine making in rural areas.

CHAPTER 2

BASIC HARVEST AND POST-HARVEST HANDLING CONSIDERATIONS FOR FRESH FRUITS AND VEGETABLES

2.1 Harvest handling

2.1.1 Maturity index for fruits and vegetables

The principles dictating at which stage of maturity a fruit or vegetable should be harvested are crucial to its subsequent storage and marketable life and quality. Post-harvest physiologists distinguish three stages in the life span of fruits and vegetables: maturation, ripening, and senescence. Maturation is indicative of the fruit being ready for harvest. At this point, the edible part of the fruit or vegetable is fully developed in size, although it may not be ready for immediate consumption. Ripening follows or overlaps maturation, rendering the produce edible, as indicated by taste. Senescence is the last stage, characterized by natural degradation of the fruit or vegetable, as in loss of texture, flavour, etc. (senescence ends at the death of the tissue of the fruit). Some typical maturity indexes are described in following sections.

Skin colour:

This factor is commonly applied to fruits, since skin colour changes as fruit ripens or matures. Some fruits exhibit no perceptible colour change during maturation, depending on the type of fruit or vegetable. Assessment of harvest maturity by skin colour depends on the judgment of the harvester, but colour charts are available for cultivars, such as apples, tomatoes, peaches, chilli peppers, etc.

Optical methods:

Light transmission properties can be used to measure the degree of maturity of fruits. These methods are based on the chlorophyll content of the fruit, which is reduced during maturation. The fruit is exposed to a bright light, which is then switched off so that the fruit is in total darkness. Next, a sensor measures the amount of light emitted from the fruit, which is proportional to its chlorophyll content and thus its maturity.

Shape:

The shape of fruit can change during maturation and can be used as a characteristic to determine harvest maturity. For instance, a banana becomes more rounded in cross-sections and less angular as it develops on the plant. Mangoes also change shape during maturation. As the mango matures on the tree the relationship between the shoulders of the fruit and the point at which the stalk is attached may change. The shoulders of immature mangoes slope away from the fruit stalk; however, on more mature mangoes the shoulders become level with the point of attachment, and with even more maturity the shoulders may be raised above this point.

Size:

Changes in the size of a crop while growing are frequently used to determine the time of harvest. For example, partially mature cobs of *Zea mays saccharata* are marketed as sweet corn, while even less mature and thus smaller cobs are marketed as baby corn. For bananas, the width of individual fingers can be used to determine harvest maturity. Usually a finger is placed midway along the bunch and its maximum width is measured with callipers; this is referred to as the calliper grade.

Aroma:

Most fruits synthesize volatile chemicals as they ripen. Such chemicals give fruit its characteristic odour and can be used to determine whether it is ripe or not. These doors may only be detectable by humans when a fruit is completely ripe, and therefore has limited use in commercial situations.

Fruit opening:

Some fruits may develop toxic compounds during ripening, such as ackee tree fruit, which contains toxic levels of hypoglycine. The fruit splits when it is fully mature, revealing black seeds on yellow arils. At this stage, it has been shown to contain minimal amounts of hypoglycine or none at all. This creates a problem in marketing; because the fruit is so mature, it will have a very short post-harvest life. Analysis of hypoglycine 'A' (hyp.) in ackee tree fruit revealed that the seed contained appreciable hyp. at all stages of maturity, at approximately 1000 ppm, while levels in the membrane mirrored those in the arils. This analysis supports earlier observations that unopened or partially opened ackee fruit should not be consumed, whereas fruit that opens naturally to over 15 mm of lobe separation poses little health hazard, provided the seed and membrane portions are removed. These observations agree with those of Brown et al. (1992) who stated that bright red, full sized ackee should never be forced open for human consumption.

Leaf changes:

Leaf quality often determines when fruits and vegetables should be harvested. In root crops, the condition of the leaves can likewise indicate the condition of the crop below ground. For example, if potatoes are to be stored, then the optimum harvest time is soon after the leaves and stems have died. If harvested earlier, the skins will be less resistant to harvesting and handling damage and more prone to storage diseases.

Abscission:

As part of the natural development of a fruit an abscission layer is formed in the pedicel. For example, in cantaloupe melons, harvesting before the abscission layer is fully developed results in inferior flavoured fruit, compared to those left on the vine for the full period.

Firmness:

A fruit may change in texture during maturation, especially during ripening when it may become rapidly softer. Excessive loss of moisture may also affect the texture of crops. These textural changes are detected by touch, and the harvester may simply be able to gently squeeze the fruit and judge whether the crop can be harvested. Today sophisticated devices have been developed to measure texture in fruits and vegetables, for example, texture analyzers and pressure testers; they are currently available for fruits and vegetables in various forms. A force is applied to the surface of the fruit, allowing the probe of the penetrometer or texturometer to penetrate the fruit flesh, which then gives a reading on firmness. Hand held pressure testers could give variable results because the basis on which they are used to measure firmness is affected by the angle at which the force is applied. Two commonly used pressure testers to measure the firmness of fruits and vegetables are the Magness-Taylor and UC Fruit Firmness testers (Figure 2.1). A more elaborate test, but not necessarily more effective, uses instruments like the Instron Universal Testing Machine. It is necessary to specify the instrument and all settings used when reporting test pressure values or attempting to set standards.

The Agricultural Code of California states that "Bartlett pears shall be considered mature if they comply with one of the following: (a) the average pressure test of not less than 10 representative pears for each commercial size in any lot does not exceed 23 lb (10.4 kg); (b) the soluble solids in a sample of juice from not less than 10 representative pears for each commercial size in any lot is not less than 13%" (Ryall and Pentzer, 1982). This Code defines minimum maturity for Bartlett pears and is presented in Table 2.1.

Table 2.1. Minimum maturity standard (expressed as minimum soluble solids required and maximum Magness-Taylor test pressure allowed) of fresh Bartlett pears for selected pear size ranges (adapted from Ryall and Pentzer, 1982).

*Pear Size**	*6.0 cm to 6.35 cm*	*≥6.35 cm*
Minimum Soluble Solids (%)	**Maximum Test Pressure (kg)**	
Below 10%	8.6	9.1
10%	9.1	9.5
11%	9.3	9.8
12%	9.5	10.0

* *Pear size expressed as maximum diameter (cm)*

Table 2.1 can be simplified by establishing a minimum tolerance level of 13% soluble solids as indicator of a pear's maturity and in this way avoid the pressure test standard control (California Pear Bulletin No. 1, 1972, California Tree Fruit Agreement, Sacramento, CA):

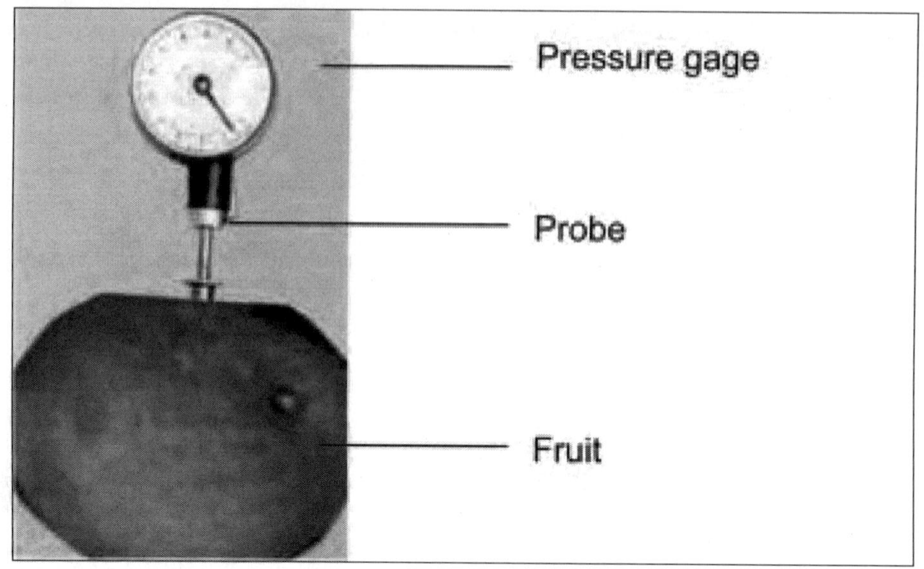

Figure 2.1 Pressure tester used to measure firmness of fruits and vegetables.

Juice content:

The juice content of many fruits increases as the fruit matures on the tree. To measure the juice content of a fruit, a representative sample of fruit is taken and then the juice extracted in a standard and specified manner. The juice volume is related to the original mass of juice, which is proportional to its maturity. The minimum values for citrus juices are presented in Table 2.2.

Table 2.2. Minimum juice values for mature citrus.

Citrus fruit	Minimum juice content (%)
Naval oranges	30
Other oranges	35
Grapefruit	35
Lemons	25
Mandarins	33
Clementines	40

Oil content and dry matter percentage:

Oil content can be used to determine the maturity of fruits, such as avocados. According to the Agricultural Code in California, avocados at the time of harvest and at any time thereafter, shall not contain in weight less than 8% oil per avocado, excluding skin and seed (Mexican or Guatemalan race cultivars). Thus, the oil content of an avocado is related to moisture content. The oil content is determined by weighing 5-10 g of avocado pulp and then extracting the oil with a solvent (e.g., benzene or petroleum ether) in a destillation column. This method has been successful for cultivars naturally high in oil content (Nagy and Shaw, 1980) .

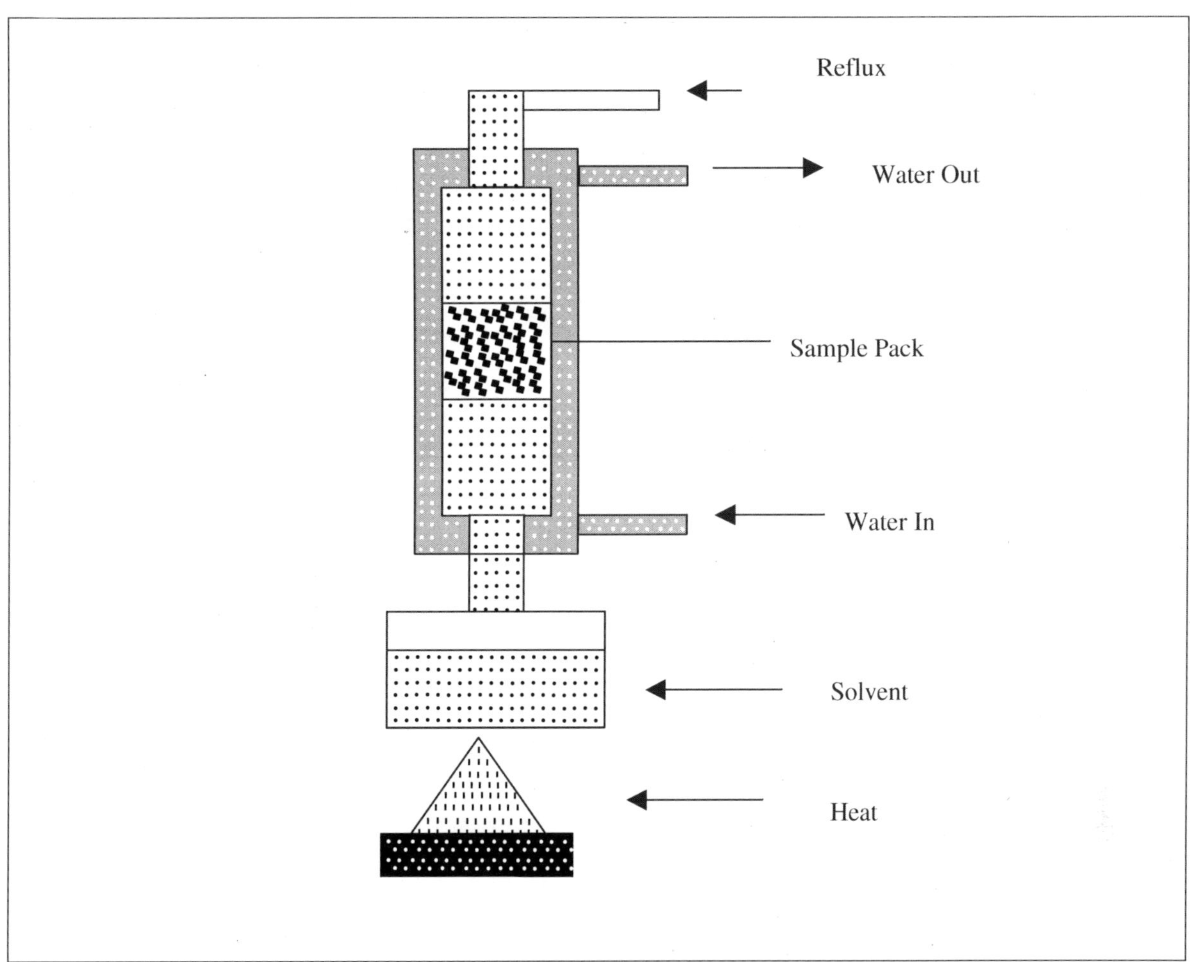

Figure 2.2 Distillation column used for oil determination.

A round flask is used for the solvent. Heat is supplied with an electric plate and water recirculated to maintain a constant temperature during the extraction process (Figure 2.2). Extraction is performed using solvents such as petroleum ether, benzene, diethyl ether, etc., a process that takes between 4-6 h. After the extraction, the oil is recovered from the flask through evaporation of the water at 105°C in an oven until constant weight is achieved.

Moisture content

During the development of avocado fruit the oil content increases and moisture content rapidly decreases (Olaeta-Coscorroza and Undurraga-Martinez, 1995). The moisture levels required to obtain good acceptability of a variety of avocados cultivated in Chile are listed in Table 2.3.

Table 2.3. Moisture content of avocado fruit cultivated in Chile.

Cultivar	Moisture content (%)
Negra de la Cruz	80.1
Bacon	77.5
Zutano	80.5
Fuerte	77.9
Edranol	78.1
Hass	73.8
Gwen	78.4
Whitesell	79.1

Sugars:
In climacteric fruits, carbohydrates accumulate during maturation in the form of starch. As the fruit ripens, starch is broken down into sugar. In non-climacteric fruits, sugar tends to accumulate during maturation. A quick method to measure the amount of sugar present in fruits is with a brix hydrometer or a refractometer. A drop of fruit juice is placed in the sample holder of the refractometer and a reading taken; this is equivalent to the total amount of soluble solids or sugar content. This factor is used in many parts of the world to specify maturity. The soluble solids content of fruit is also determined by shining light on the fruit or vegetable and measuring the amount transmitted. This is a laboratory technique however and might not be suitable for village level production.

Starch content:
Measurement of starch content is a reliable technique used to determine maturity in pear cultivars. The method involves cutting the fruit in two and dipping the cut pieces into a solution containing 4% potassium iodide and 1% iodine. The cut surfaces stain to a blue-black colour in places where starch is present. Starch converts into sugar as harvest time approaches. Harvest begins when the samples show that 65-70% of the cut surfaces have turned blue-black.

Acidity:
In many fruits, the acidity changes during maturation and ripening, and in the case of citrus and other fruits, acidity reduces progressively as the fruit matures on the tree.

Taking samples of such fruits, and extracting the juice and titrating it against a standard alkaline solution, gives a measure that can be related to optimum times of harvest. Normally, acidity is not taken as a measurement of fruit maturity by itself but in relation to soluble solids, giving what is termed the brix: acid ratio. Sanchez et al. (1996) studied the effect of inducing maturity in banana (*Musa sp (L.), AAB*) "Silk" fruits with 2-chloroethyl phosphoric acid ("ethephon"), in some trials in Venezuela. Four treatments (0, 1000, 3000, and 5000 ppm) were applied. The results obtained revealed that the "ethephon" treatments increased the acidity and total soluble solids. The sucrose formation accelerated while the pH was not affected significantly. On the other hand, the relationship of the Brix/acidity ratio was increased according to the "ethephon" dose, as presented in Table 2.4.

Table 2.4. Effect of ethephon on the maturity index (°Brix/acidity ratio) of banana (manzano) Silk fruits.

		Ethephon doses (ppm)			
Stage of maturity	**Days**	**0**	**1000**	**3000**	**5000**
Green	1	29.35[a]	23.99[a]	20.59[b]	19.31[b]
Slightly ripen	3	33.27[c]	33.53[c]	58.29[a]	46.27[b]
Slightly ripen	5	51.15[b]	66.44[a]	63.01[b]	57.00[c]
Slightly ripen	7	60.69[a]	69.35[a]	64.31[a]	68.35[a]
Ripen	9	53.27[a]	57.36[a]	54.67[a]	55.42[a]
Variation (%)		81.50	139.10	165.52	187.00
(Full ripen)					

Means with different letters in a row are significantly different at $p < 0.05$ (Tukey Test).

Specific gravity:

Specific gravity is the relative gravity, or weight of solids or liquids, compared to pure distilled water at 62°F (16.7°C), which is considered unity. Specific gravity is obtained by comparing the weights of equal bulks of other bodies with the weight of water. In practice, the fruit or vegetable is weighed in air, then in pure water. The weight in air divided by the weight in water gives the specific gravity. This will ensure a reliable measure of fruit maturity. As a fruit matures its specific gravity increases. This parameter is rarely used in practice to determine time of harvest, but could be used in cases where development of a suitable sampling technique is possible. It is used however to grade crops according to different maturities at post-harvest. This is done by placing the fruit in a tank of water, wherein those that float are less mature than those that sink.

2.1.2 Harvesting containers

Harvesting containers must be easy to handle for workers picking fruits and vegetables in the field. Many crops are harvested into bags. Harvesting bags with shoulder or waist slings can be used for fruits with firm skins, like citrus fruits and avocados. These containers are made from a variety of materials such as paper, polyethylene film, sisal, hessian or woven polyethylene and are relatively cheap but give little protection to the crop against handling and transport damage. Sacks are commonly used for crops such as potatoes, onions, cassava, and pumpkins. Other types of field harvest containers include baskets, buckets, carts, and plastic crates (Figure 2.3). For high risk products, woven baskets and sacks are not recommended because of the risk of contamination.

Figure 2.3 Agricultural apple baskets, pear and corncob carriers.

2.1.3 Tools for harvesting

Depending on the type of fruit or vegetable, several devices are employed to harvest produce. Commonly used tools for fruit and vegetable harvesting are secateurs or knives, and hand held or pole mounted picking shears. When fruits or vegetables are difficult to catch, such as mangoes or avocados, a cushioning material is placed around the tree to prevent damage to the fruit when dropping from high trees. Harvesting bags with shoulder or waist slings can be used for fruits with firm skins, like citrus and avocados. They are easy to carry and leave both hands free. The contents of the bag are emptied through the bottom into a field container without tipping the bag. Plastic buckets are suitable containers for harvesting fruits that are easily crushed, such as tomatoes. These containers should be smooth without any sharp edges that could damage the produce. Commercial growers use bulk bins with a capacity of 250-500 kg, in which crops such as apples and cabbages are placed, and sent to large-scale packinghouses for selection, grading, and packing.

2.1.4 Packing in the field and transport to packinghouse

Berries picked for the fresh market (except blueberries and cranberries) are often mechanically harvested and usually packed into shipping containers. Careful harvesting, handling, and transporting of fruits and vegetables to packinghouses are necessary to preserve product quality.

Polyethylene bags:

Clear polyethylene bags are used to pack banana bunches in the field, which are then transported to the packinghouse by means of mechanical cableways running through the banana plantation. This technique of packaging and transporting bananas reduces damage to the fruit caused by improper handling.

Plastic field boxes:

These types of boxes are usually made of polyvinyl chloride, polypropylene, or polyethylene. They are durable and can last many years. Many are designed in such a way that they can nest inside each other when empty to facilitate transport, and can stack one on top of the other without crushing the fruit when full (Figure 2.4).

Figure 2.4 Plastic field boxes with nest/stack design.

Wooden field boxes:

These boxes are made of thin pieces of wood bound together with wire. They come in two sizes: the bushel box with a volume of 2200 in^3 (36052 cm^3) and the half-bushel box. They are advantageous because they can be packed flat and are inexpensive, and thus could be non-returnable. They have the disadvantage of providing little protection from mechanical damage to the produce during transport. Rigid wooden boxes of different capacities are commonly used to transport produce to the packinghouse or to market. (Figure 2.5).

Figure 2.5 Typical wooden crate holding fresh tomatoes.

Bulk bins:

Bulk bins of 200-500 kg capacity are used for harvesting fresh fruits and vegetables. These bins are much more economical than the field boxes, both in terms of fruit carried per unit volume and durability, as well as in providing better protection to the product during transport to the packinghouse. They are made of wood and plastic materials. Dimensions for these bins in the United States are 48 X 40 in, and 120 X 100 cm in metric system countries. Approximate depth of bulk bins depends on the type of fruit or vegetable being transported (Table 2.5)

Table 2.5. Approximate depth of bulk bins.

Commodity	Depth (cm)
Citrus	70
Pears, apples	50
Stone fruits	50
Tomatoes	40

2.2 Post-harvest handling

2.2.1 Curing of roots, tubers, and bulb crops

When roots and tubers are to be stored for long periods, curing is necessary to extend the shelf life. The curing process involves the application of high temperatures and high relative humidity to the roots and tubers for long periods, in order to heal the skins wounded during harvesting. With this process a new protected layer of cells is formed. Initially the curing process is expensive, but in the long run, it is worthwhile. The conditions for curing roots and tubers are presented in Table 2.6.

Table 2.6. Conditions for curing roots and tubers.

Commodity	**Temperature (°C)**	**Relative Humidity (%)**	**Storage time (days)**
Potato	15-20	90-95	5-10
Sweet potato	30-32	85-90	4-7
Yams	32-40	90-100	1-4
Cassava	30-40	90-95	2-5

Source: FAO (1995a)

Curing can be accomplished in the field or in curing structures conditioned for that purpose. Commodities such as yams can be cured in the field by piling them in a partially shaded area. Cut grass or straw can serve as insulating material while covering the pile with canvas, burlap, or woven grass matting. This covering will provide sufficient heat to reach high temperatures and high relative humidity. The stack can be left in this state for up to four days.

Onions and garlic can be cured in the field in windrows or after being packed into large fibre or net sacks. Modern curing systems have been implemented in housing conditioned with fans and heaters to produce the heat necessary for high temperatures and high relative humidity, as illustrated below:

The fans are used to redistribute the heat to the lower part of the room where the produce is stored. Bulk bins are stacked with a gap of 10 to 15 cm between rows to allow adequate air passage. The system shown in Figure 2.6 can be used for curing onions; an exhaust opening near the ceiling must be provided for air recirculation. Care should be taken to prevent over-dryness of the onion bulbs.

When extreme conditions in the field exist, such as heavy rain or flooded terrain, and curing facilities are not available, a temporary tent must be constructed from large tarpaulins or plastic sheets to cure the onions and avoid heavy loss. Heated air is forced into a hollow area at the centre of the produce-filled bins. Several fans are used to recirculate the warm air through the onions while curing.

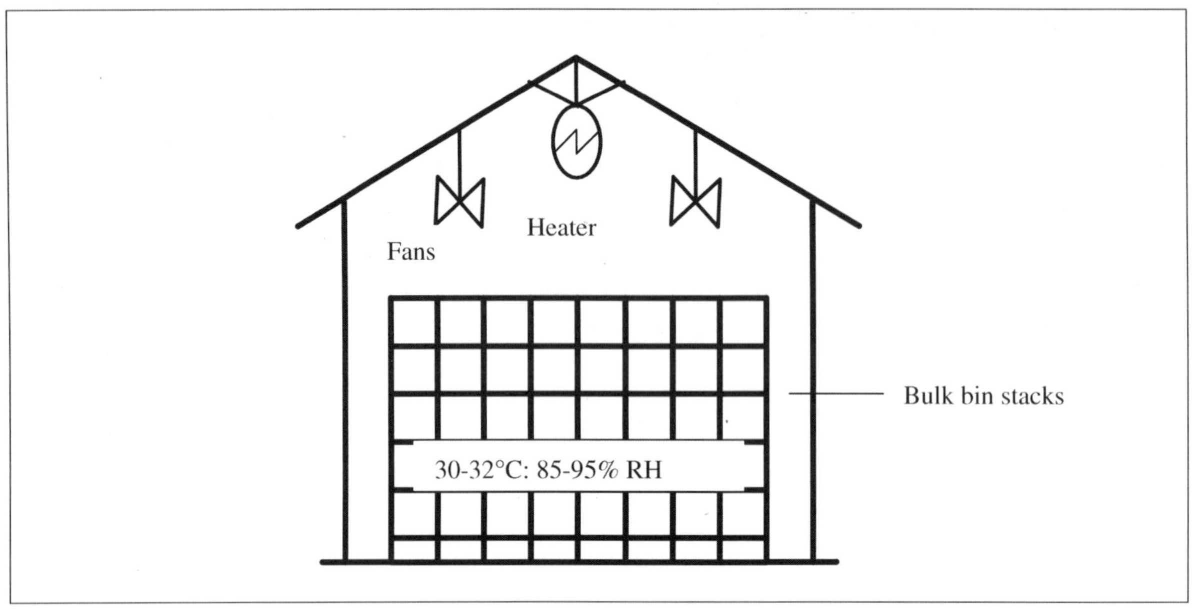

Figure 2.6 Typical curing houses for roots and tubers.

2.2.2 Operations prior to packaging

Fruits and vegetables are subjected to preliminary treatments designed to improve appearance and maintain quality. These preparatory treatments include cleaning, disinfection, waxing, and adding of colour (some includes brand name stamping on individual fruits).

Cleaning:

Most produce receives various chemical treatments such as spraying of insecticides and pesticides in the field. Most of these chemicals are poisonous to humans, even in small concentrations. Therefore, all traces of chemicals must be removed from produce before packing. As illustrated in Figure 2.7, the fruit or vegetable passes over rotary brushes where it is rotated and transported to the washing machine and exposed to the cleaning process from all sides:

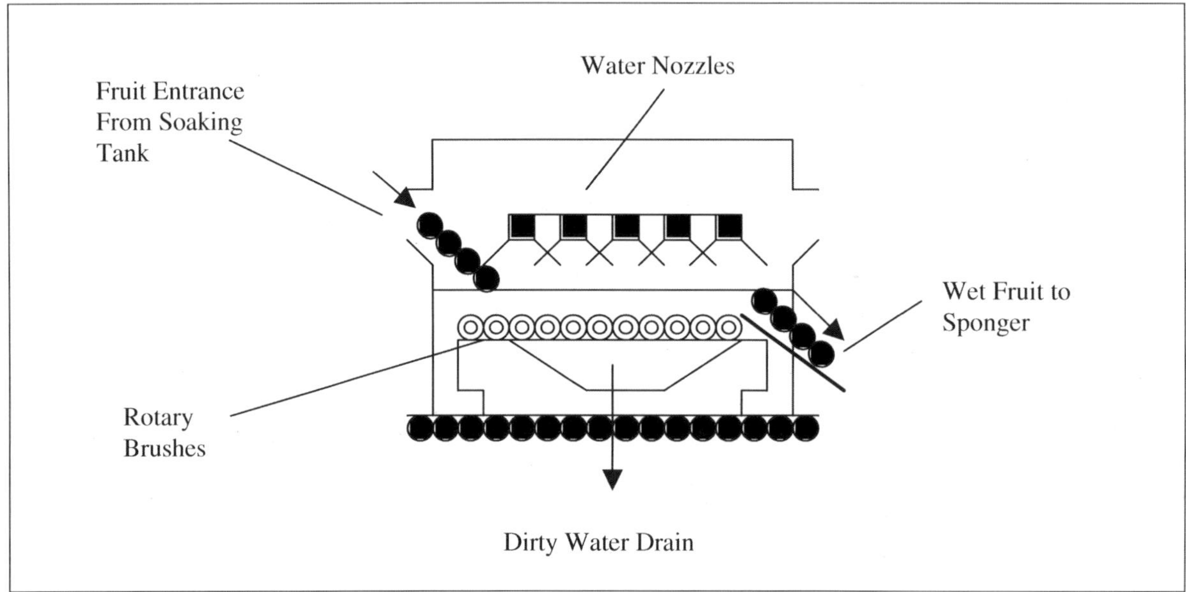

Figure 2.7 Typical produce washing machine.

From the washing machine, the fruit passes onto a set of rotary sponge rollers (similar to the rotary brushes). The rotary sponges remove most of the water on the fruit as it is rotated and transported through the sponger.

Disinfection:
After washing fruits and vegetables, disinfectant agents are added to the soaking tank to avoid propagation of diseases among consecutive batches of produce. In a soaking tank, a typical solution for citrus fruit includes a mixture of various chemicals at specific concentration, pH, and temperature, as well as detergents and water softeners. Sodium-ortho-phenyl-phenate (SOPP) is an effective citrus disinfectant, but requires precise control of conditions in the tank. Concentrations must be kept between 0.05 and 0.15%, with pH at 11.8 and temperature in the range of 43-48°C. Recommended soaking time is 3-5 minutes. Deviation from these recommendations may have disastrous effects on the produce, since the solution will be ineffective if the temperature or concentration is too low (Peleg, 1985). Low concentrations of chlorine solution are also used as disinfectant for many vegetables. The advantage of this solution is that it does not leave a chemical residue on the product.

Artificial waxing:
Artificial wax is applied to produce to replace the natural wax lost during washing of fruits or vegetables. This adds a bright sheen to the product. The function of artificial waxing of produce is summarized below:

- Provides a protective coating over entire surface.
- Seals small cracks and dents in the rind or skin.
- Seals off stem scars or base of petiole.
- Reduces moisture loss.
- Permits natural respiration.
- Extends shelf life.
- Enhances sales appeal.

Brand name application:
Some distributors use ink or stickers to stamp a brand name or logo on each individual fruit. Ink is not permissible in some countries (e.g., Japan), but stickers are acceptable. Automatic machines for dispensing and applying pressure sensitive paper stickers are readily available. The advantage of stickers is that they can be easily peeled off.

2.2.3 Packaging

According to Wills et al. (1989), modern packaging must comply with the following requirements:

a) The package must have sufficient mechanical strength to protect the contents during handling, transport, and stacking.

b) The packaging material must be free of chemical substances that could transfer to the produce and become toxic to man.

c) The package must meet handling and marketing requirements in terms of weight, size, and shape.

d) The package should allow rapid cooling of the contents. Furthermore, the permeability of plastic films to respiratory gases could also be important.

e) Mechanical strength of the package should be largely unaffected by moisture content (when wet) or high humidity conditions.
f) The security of the package or ease of opening and closing might be important in some marketing situations.
g) The package must either exclude light or be transparent.
h) The package should be appropriate for retail presentations.
i) The package should be designed for ease of disposal, re-use, or recycling.
j) Cost of the package in relation to value and the extent of contents protection required should be as low as possible.

Classification of packaging:
Packages can be classified as follows:

- Flexible sacks; made of plastic jute, such as bags (small sacks) and nets (made of open mesh)
- Wooden crates
- Cartons (fibreboard boxes)
- Plastic crates
- Pallet boxes and shipping containers
- Baskets made of woven strips of leaves, bamboo, plastic, etc.

Uses for above packages:

Nets are only suitable for hard produce such as coconuts and root crops (potatoes, onions, yams).

Wooden crates are typically wire bound crates used for citrus fruits and potatoes, or wooden field crates used for softer produce like tomatoes. Wooden crates are resistant to weather and more efficient for large fruits, such as watermelons and other melons, and generally have good ventilation. Disadvantages are that rough surfaces and splinters can cause damage to the produce, they can retain undesirable odours when painted, and raw wood can easily become contaminated with moulds.

Fibreboard boxes are used for tomato, cucumber, and ginger transport. They are easy to handle, light weight, come in different sizes, and come in a variety of colours that can make produce more attractive to consumers. They have some disadvantages, such as the effect of high humidity, which can weaken the box; neither are they waterproof, so wet products would need to be dried before packaging. These boxes are often of lower strength compared to wooden or plastic crates, although multiple thickness trays are very widely used. They can come flat packed with ventilation holes and grab handles, making a cheap attractive alternative that is very popular. Care should be taken that holes on the surface (top and sides) of the box allow adequate ventilation for the produce and prevent heat generation, which can cause rapid product deterioration.

Plastic crates are expensive but last longer than wooden or carton crates.
They are easy to clean due to their smooth surface and are hard in strength, giving protection to products. Plastic crates (Figure 2.8) can be used many times, reducing the cost of transport.

They are available in different sizes and colours and are resistant to adverse weather conditions. However, plastic crates can damage some soft produce due to their hard surfaces, thus liners are recommended when using such crates.

Pallet boxes are very efficient for transporting produce from the field to the packinghouse or for handling produce in the packinghouse. Pallet boxes have a standard floor size (1200 x 1000 mm) and depending on the commodity have standard heights. Advantages of the pallet box are that it reduces the labour and cost of loading, filling, and unloading; reduces space for storage; and increases speed of mechanical harvest. The major disadvantage is that the return volume of most pallet boxes is the same as the full load. Higher investment is also required for the forklift truck, trailer, and handling systems to empty the boxes. They are not affordable to small producers because of high, initial capital investment.

Figure 2.8 Typical plastic crate holding fresh oranges.

2.2.4 Cooling methods and temperatures

Several methods of cooling are applied to produce after harvesting to extend shelf life and maintain a fresh-like quality. Some of the low temperature treatments are unsuitable for simple rural or village treatment but are included for consideration as follows:

2.2.4.1 Precooling

Fruit is precooled when its temperature is reduced from 3 to 6°C (5 to 10°F) and is cool enough for safe transport. Precooling may be done with cold air, cold water (hydrocooling), direct contact with ice, or by evaporation of water from the product under a partial vacuum (vacuum cooling). A combination of cooled air and water in the form of a mist called hyraircooling is an innovation in cooling of vegetables.

2.2.4.2 Air precooling

Precooling of fruits with cold air is the most common practice. It can be done in refrigerator cars, storage rooms, tunnels, or forced air-coolers (air is forced to pass through the container via baffles and pressure differences).

2.2.4.3 Icing

Ice is commonly added to boxes of produce by placing a layer of crushed ice directly on the top of the crop. An ice slurry can be applied in the following proportion: 60% finely crushed ice, 40% water, and 0.1% sodium chloride to lower the melting point. The water to ice ratio may vary from 1:1 to 1:4.

2.2.4.4 Room cooling

This method involves placing the crop in cold storage. The type of room used may vary, but generally consists of a refrigeration unit in which cold air is passed through a fan. The circulation may be such that air is blown across the top of the room and falls through the crop by convection. The main advantage is cost because no specific facility is required.

2.2.4.5 Forced air-cooling

The principle behind this type of precooling is to place the crop into a room where cold air is directed through the crop after flowing over various refrigerated metal coils or pipes. Forced air-cooling systems blow air at a high velocity leading to desiccation of the crop. To minimize this effect, various methods of humidifying the cooling air have been designed such as blowing the air through cold water sprays.

2.2.4.6 Hydrocooling

The transmission of heat from a solid to a liquid is faster than the transmission of heat from a solid to a gas. Therefore, cooling of crops with cooled water can occur quickly and results in zero loss of weight. To achieve high performance, the crop is submerged in cold water, which is constantly circulated through a heat exchanger. When crops are transported around the packhouse in water, the transport can incorporate a hydrocooler. This system has the advantage wherein the speed of the conveyer can be adjusted to the time required to cool the produce. Hydrocooling has a further advantage over other precooling methods in that it can help clean the produce. Chlorinated water can be used to avoid spoilage of the crop. Hydrocooling is commonly used for vegetables, such as asparagus, celery, sweet corn, radishes, and carrots, but it is seldom used for fruits.

2.2.4.7 Vacuum cooling

Cooling in this case is achieved with the latent heat of vaporization rather than conduction. At normal air pressure (760 mmHg) water will boil at 100°C. As air pressure is reduced so is the boiling point of water, and at 4.6 mmHg water boils at 0°C. For every 5 or 6°C reduction in temperature, under these conditions, the crop loses about 1% of its weight (Barger, 1961). This weight loss may be minimized by spraying the produce with water either before enclosing it in the vacuum chamber or towards the end of the vacuum cooling operation (hydrovacuum cooling). The speed and effectiveness of cooling is related to the ratio between the mass of the crop and its surface area. This method is particularly suitable for leaf crops such as lettuce. Crops like tomatoes having a relatively thick wax cuticle are not suitable for vacuum cooling.

2.2.4.8 Recommended minimum temperature to increase storage time

There is no ideal storage for all fruits and vegetables, because their response to reduced temperatures varies widely. The importance of factors such as mould growth and chilling injuries must be taken into account, as well as the required length of storage (Wills et al., 1989). Storage temperature for fruits and vegetables can range from –1 to 13°C, depending on their perishability. Extremely perishable fruits such as apricots, berries, cherries, figs, watermelons can be stored at –1 to 4°C for 1-5 weeks; less perishable fruits such as mandarin, nectarine, ripe or green pineapple can be stored at 5-9°C for 2–5 weeks; bananas at 10°C for 1-2 weeks and green bananas at 13°C for 1-2 weeks. Highly perishable vegetables can be stored up to 4 weeks such as asparagus, beans, broccoli, and Brussels sprouts at –1-4°C for 1-4 weeks; cauliflower at 5-9°C for 2-4 weeks. Green tomato is less perishable and can be stored at 10°C for 3-6 weeks and non-perishable vegetables such as carrots, onions, potatoes and parsnips can be stored at 5-9°C for 12-28 weeks. Similarly, sweet potatoes can be stored at 10°C for 16-24 weeks. The storage life of produce is highly variable and related to the respiration rate; there is an inverse relation between respiration rate and storage life in that produce with low respiration generally keeps longer.

For example, the respiration rate of a very perishable fruit like ripe banana is 200 mL $CO_2.kg^{-1}h^{-1}$ at 15°C, compared to a non-perishable fruit such as apple, which has a respiration rate of 25 mL $CO_2.kg^{-1}h^{-1}$ at 15°C.

2.2.4.9 High temperatures

Exposure of fruits and vegetables to high temperatures during post-harvest reduces their storage or marketable life. This is because as living material, their metabolic rate is normally higher with higher temperatures. High temperature treatments are beneficial in curing root crops, drying bulb crops, and controlling diseases and pests in some fruits. Many fruits are exposed to high temperatures in combination with ethylene (or another suitable gas) to initiate or improve ripening or skin colour.

2.2.5 Storage

The marketable life of most fresh vegetables can be extended by prompt storage in an environment that maintains product quality. The desired environment can be obtained in facilities where temperature, air circulation, relative humidity, and sometimes atmosphere composition can be controlled. Storage rooms can be grouped accordingly as those requiring refrigeration and those that do not. Storage rooms and methods not requiring refrigeration include: *in situ*, sand, coir, pits, clamps, windbreaks, cellars, barns, evaporative cooling, and night ventilation:

In situ. This method of storing fruits and vegetables involves delaying the harvest until the crop is required. It can be used in some cases with root crops, such as cassava, but means that the land on which the crop was grown will remain occupied and a new crop cannot be planted. In colder climates, the crop may be exposed to freezing and chilling injury.

Sand or coir: This storage technique is used in countries like India to store potatoes for longer periods of time, which involves covering the commodity under ground with sand.

Pits or trenches are dug at the edges of the field where the crop has been grown. Usually pits are placed at the highest point in the field, especially in regions of high rainfall. The pit or trench is lined with straw or other organic material and filled with the crop being stored, then covered with a layer of organic material followed by a layer of soil. Holes are created with straw at the top to allow for air ventilation, as lack of ventilation may cause problems with rotting of the crop.

Clamps. This has been a traditional method for storing potatoes in some parts of the world, such as Great Britain. A common design uses an area of land at the side of the field. The width of the clamp is about 1 to 2.5 m. The dimensions are marked out and the potatoes piled on the ground in an elongated conical heap. Sometimes straw is laid on the soil before the potatoes. The central height of the heap depends on its angle of repose, which is about one third the width of the clump. At the top, straw is bent over the ridge so that rain will tend to run off the structure. Straw thickness should be from 15-25 cm when compressed. After two weeks, the clamp is covered with soil to a depth of 15-20 cm, but this may vary depending on the climate.

Windbreaks are constructed by driving wooden stakes into the ground in two parallel rows about 1 m apart. A wooden platform is built between the stakes about 30 cm from the ground, often made from wooden boxes. Chicken wire is affixed between the stakes and across both ends of the windbreak. This method is used in Britain to store onions (Thompson, 1996).

Cellars. These underground or partly underground rooms are often beneath a house. This location has good insulation, providing cooling in warm ambient conditions and protection from excessively low temperatures in cold climates. Cellars have traditionally been used at domestic scale in Britain to store apples, cabbages, onions, and potatoes during winter.

Barns. A barn is a farm building for sheltering, processing, and storing agricultural products, animals, and implements. Although there is no precise scale or measure for the type or size of the building, the term barn is usually reserved for the largest or most important structure on any particular farm. Smaller or minor agricultural buildings are often labelled sheds or outbuildings and are normally used to house smaller implements or activities.

Evaporative cooling. When water evaporates from the liquid phase into the vapour phase energy is required. This principle can be used to cool stores by first passing the air introduced into the storage room through a pad of water. The degree of cooling depends on the original humidity of the air and the efficiency of the evaporating surface. If the ambient air has low humidity and is humidified to around 100% RH, then a large reduction in temperature will be achieved. This can provide cool moist conditions during storage.

Night ventilation. In hot climates, the variation between day and night temperatures can be used to keep stores cool. The storage room should be well insulated when the crop is placed inside. A fan is built into the store room, which is switched on when the outside temperature at night becomes lower than the temperature within. The fan switches off when the temperatures equalize.

The fan is controlled by a differential thermostat, which constantly compares the outside air temperature with the internal storage temperature. This method is used to store bulk onions.

Controlled atmospheres are made of gastight chambers with insulated walls, ceiling, and floor. They are increasingly common for fruit storage at larger scale. Depending on the species and variety, various blends of O_2, CO_2, and N_2 are required. Low content O_2 atmospheres (0.8 to 1.5%), called ULO (Ultra –Low Oxygen) atmospheres, are used for fruits with long storage lives (e.g., apples).

2.2.6 Pest control and decay

Crops may be immersed in hot water before storage or marketing to control disease. A common disease of fruits known as anthracnose, caused by the infection of fungus *Colletotrychum spp.* can be successfully controlled in this way. Combining appropriate doses of fungicides with hot water is often effective in controlling disease in fruits after harvesting. Recommended conditions for hot water treatment for controlling diseases in fruits are shown in Table 2.7:

Table 2.7. Recommended conditions for hot water and fungicide treatments.

Water Temperature (°C)	Dipping time (min)	Fungicide (ppm)	Fungus	Fruit	Decay
55-53	5	benomyl 500 flusilazole 100	*Colletotrychum)*	Mango	Controlled
46-55	3	0	None	Blueberry	90% reduced
90	0.03	0	None	Sweet potato	Delay
39	0.5	benomyl 500 dichloran 400	*Rizopus rot*	Stone fruit	Controlled
-	Few seconds	250-500 thiophanate	*Botryodiplodia theobromate*	Banana	Controlled

Source: Thompson (1998)

Fruit and vegetable decay is also caused by storage conditions. Too low temperatures can cause injury during refrigeration of fruits and vegetables. High temperatures can cause softening of tissues and promote bacterial diseases. The damage that microorganisms inflict on fresh fruits and vegetables is mainly in the physical loss of edible matter, which may be partial or total.

CHAPTER 3

GENERAL CONSIDERATIONS FOR PRESERVATION OF FRUITS AND VEGETABLES

3.1 Water Activity (a_w) concept and its role in food preservation

3.1.1 a_w concept

The concept of a_w has been very useful in food preservation and on that basis many processes could be successfully adapted and new products designed. Water has been called the universal solvent as it is a requirement for growth, metabolism, and support of many chemical reactions occurring in food products. Free water in fruit or vegetables is the water available for chemical reactions, to support microbial growth, and to act as a transporting medium for compounds. In the bound state, water is not available to participate in these reactions as it is bound by water soluble compounds such as sugar, salt, gums, etc. (osmotic binding), and by the surface effect of the substrate (matrix binding). These water-binding effects reduce the vapour pressure of the food substrate according to Raoult's Law. Comparing this vapour pressure with that of pure water (at the same temperature) results in a ratio called water activity (a_w). Pure water has an a_w of 1, one molal solution of sugar – 0.98, and one molal solution of sodium chloride – 0.9669. A saturated solution of sodium chloride has a water activity of 0.755. This same NaCl solution in a closed container will develop an equilibrium relative humidity (ERH) in a head space of 75.5%. A relationship therefore exists between ERH and a_w since both are based on vapour pressure.

$$a_{w} = \frac{ERH}{100}$$

The ERH of a food product is defined as the relative humidity of the air surrounding the food at which the product neither gains nor loses its natural moisture and is in equilibrium with the environment.

3.1.2 Microorganisms vs. a_w value

The definition of moisture conditions in which pathogenic or spoilage microorganisms cannot grow is of paramount importance to food preservation. It is well known that each microorganism has a critical a_w below which growth cannot occur. For instance, pathogenic microorganisms cannot grow at $a_w < 0.86$; yeasts and moulds are more tolerant and usually no growth occurs at $a_w < 0.62$. The so-called intermediate moisture foods (IMF) have a_w values in the range of 0.65-0.90 (Figure 3.1).

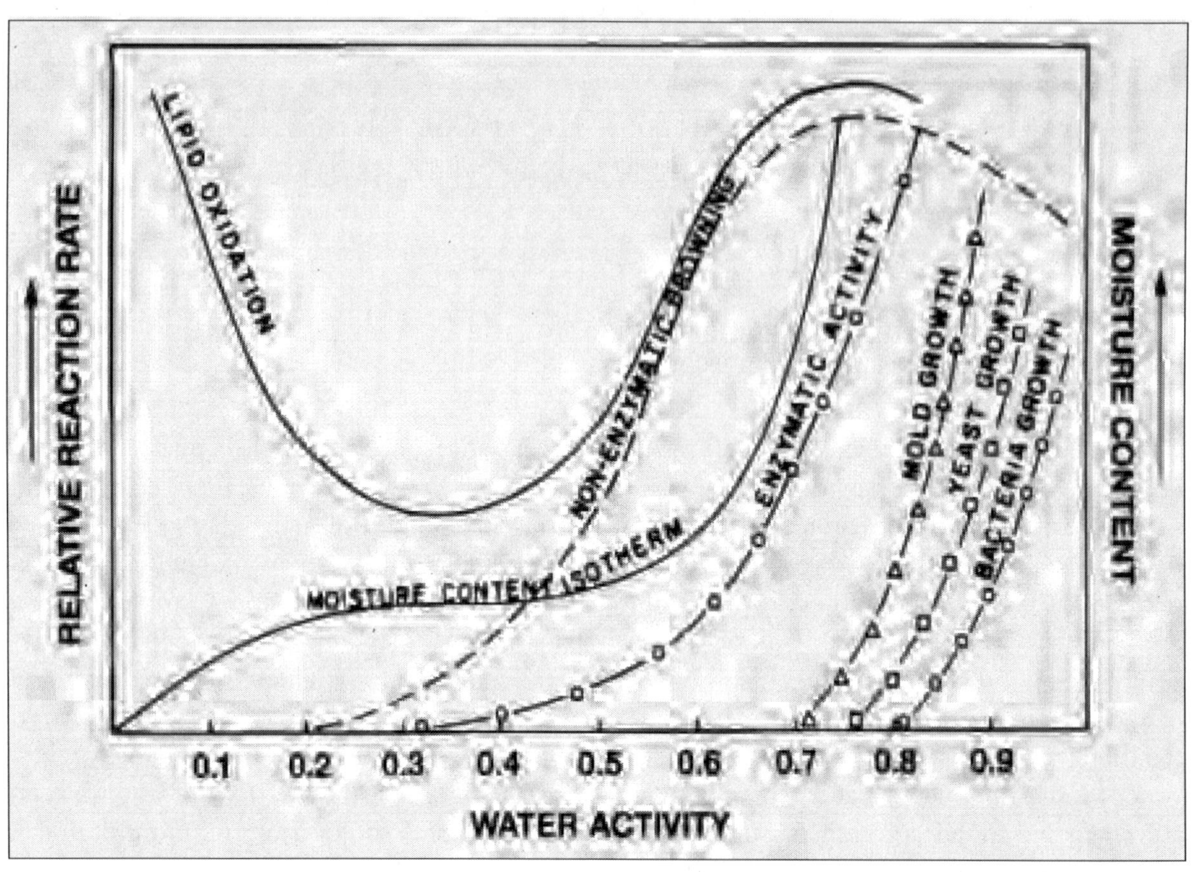

Figure 3.1 Relationship of food deterioration rate as a function of water activity.

3.1.3 Enzymatic and chemical changes related to a_w values

The relationship between enzymatic and chemical changes in foods as a function of water activity is illustrated in Figure 3.1. With a_w at 0.3, the product is most stable with respect to lipid oxidation, non-enzymatic browning, enzyme activity, and of course, the various microbial parameters. As a_w increases toward the right, the probability of the food product deteriorating increases.

According to Rahman and Labuza (1999), enzyme–catalyzed reactions can occur in foods with relatively low water contents. The authors summarized two features of these results as follows:

1. The rate of hydrolysis increases with increased water activity but is extremely slow with very low activity.
2. For each instance of water activity there appears to be a maximum amount of hydrolysis, which also increases with water content.

The apparent cessation of the reaction at low moisture cannot be due to the irreversible inactivation of the enzyme, because upon humidification to a higher water activity, hydrolysis resumes at a rate characteristic of the newly attained water activity. Rahman and Labuza(1999) reported the investigation of a model system consisting of avicel, sucrose, and invertase and found that the reaction velocity increased with water activity. Complete conversion of the substrate was observed for water activities greater than or equal to 0.75. For water activities below 0.75, the reaction continued with 100% hydrolysis. In solid media, water activity can affect reactions in two ways: lack of reactant mobility and alternation of active conformation of the substrate and enzymatic protein. The effects of varying the enzyme-to-substrate ratios on reaction velocity and the effect of water activity on the activation energy for the reaction could not be explained by a simple diffusional model, but required postulates that were more complex:

1. The diffusional resistance is localized in a shell adjacent to the enzyme.
2. At low water activity, the reduced hydration produces conformational changes in the enzyme, affecting its catalytic activity.

The relationship between water content and water activity is complex. An increase in a_w is usually accompanied by an increase in water content, but in a non-linear fashion. This relationship between water activity and moisture content at a given temperature is called the moisture sorption isotherm. These curves are determined experimentally and constitute the fingerprint of a food system.

3.1.4 Recommended equipment for measuring a_w

Many methods and instruments are available for laboratory measurement of water activity in foods. Methods are based on the colligative properties of solutions. Water activity can be estimated by measuring the following:

- Vapour pressure
- Osmotic pressure
- Freezing point depression of a liquid
- Equilibrium relative humidity of a liquid or solid
- Boiling point elevation
- Dew point and wet bulb depression
- Suction potential, or by using the isopiestic method
- Bithermal equilibrium
- Electric hygrometers
- Hair hygrometers

3.1.4.1 Vapour pressure

Water activity is expressed as the ratio of the partial pressure of water in a food to the vapour pressure of pure water with the same temperature as the food. Thus, measuring the vapour pressure of water in a food system is the most direct measure of a_w. The food sample measured is allowed to equilibrate, and measurement is taken by using a manometer or transducer device as depicted in Figure 3.2. This method can be affected by sample size, equilibration time, temperature, and volume. This method is not suitable for biological materials with active respiration or materials containing large amounts of volatiles.

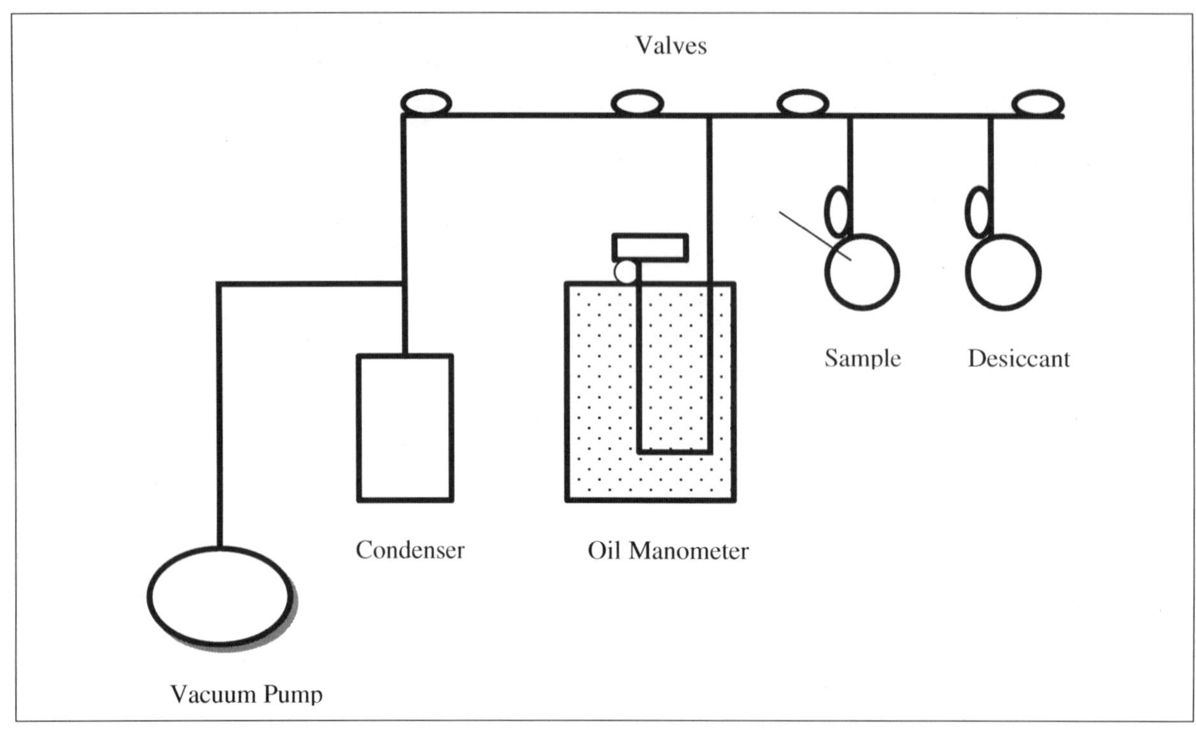

Figure 3.2 Vapour pressure manometer.
(Adapted from Barbosa-Cánovas and Vega-Mercado, 1996)

3.1.4.2 Freezing point depression and freezing point elevation

This method is accurate for liquids in the high water activity range but is not suitable for solid foods (Barbosa-Cánovas and Vega-Mercado, 1996). The water activity can be estimated using the following two expressions:

Freezing point depression:

$$-\log a_w = 0.004207\ \Delta T_f + 2.1\ E\text{-}6\ \Delta T^2{}_f \qquad (1)$$

where ΔT_f is the depression in the freezing temperature of water

Boiling point elevation:

$$-\log a_w = 0.01526\ \Delta T_b - 4.862\ E\text{-}5\ \Delta T^2{}_b \qquad (2)$$

where ΔT_b is the elevation in the boiling temperature of water.

3.1.4.3 Osmotic pressure

Water activity can be related to the osmotic pressure (π) of a solution with the following equation:

$$\pi = RT/V_w \ln(a_w) \qquad (3)$$

where V_w is the molar volume of water in solution, R the universal gas constant, and T the absolute temperature. Osmotic pressure is defined as the mechanical pressure needed to prevent a net flow of solvent across a semi-permeable membrane. For an ideal solution, Equation (3) can be redefined as:

$$\pi = RT/V_w \ln(X_w) \qquad (4)$$

where X_w is the molar fraction of water in the solution. For non-ideal solutions, the osmotic pressure expression can be rewritten as:

$$\pi = RT\phi v m_b(m_w V_w) \qquad (5)$$

where ν is the number of moles of ions formed from one mole of electrolyte, m_w and m_b are the molar concentrations of water and the solute, respectively, and φ the osmotic coefficient, defined as:

$$\phi = -m_w \ln(a_w) / \nu m_b \qquad (6)$$

3.1.4.4 Dew point hygrometer

Vapour pressure can be determined from the dew point of an air-water mixture. The temperature at which the dew point occurs is determined by observing condensation on a smooth, cool surface such as a mirror. This temperature can be related to vapour pressure using a psychrometric chart. The formation of dew is detected photoelectrically, as illustrated in the diagram below:

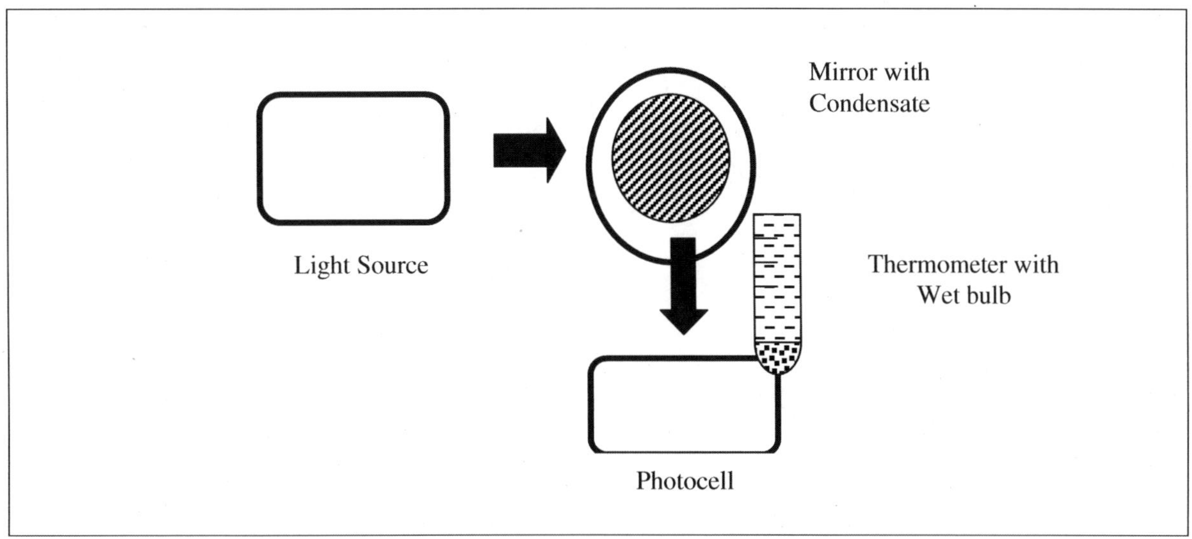

Figure 3.3 Dew point determination of water activity.
(Adapted from Barbosa-Cánovas and Vega-Mercado, 1996)

3.1.4.5 Thermocouple Psychrometer

Water activity measurement is based on wet bulb temperature depression. A thermocouple is placed in the chamber where the sample is equilibrated. Water is then sprayed over the thermocouple before it is allowed to evaporate, causing a decrease in temperature. The drop in temperature is related to the rate of water evaporation from the surface of the thermocouple, which is a function of the relative humidity in equilibrium with the sample.

3.1.4.6 Isopiestic method

The isopiestic method consists of equilibrating both a sample and a reference material in an evacuated desiccator until equilibrium is reached at 25°C. The moisture content of the reference material is then determined and the a_w obtained from the sorption isotherm. Since the sample was in equilibrium with the reference material, the a_w of both is the same.

3.1.4.7 Electric hygrometers

Most hygrometers are electrical wires coated with hygroscopic salts or sulfonated polystyrene gel in which conductance or capacitance changes as the coating absorbs moisture from the sample. The major disadvantage of this type of hygrometer is the tendency of the hygroscopic salt to become contaminated with polar compounds, resulting in erroneous a_w determinations.

3.1.4.8 Hair hygrometers

Hair hygrometers are based on the stretching of a fibre when exposed to high water activity. They are less sensitive than other instruments at lower levels of activity ($<0.03\ a_w$) and the principal disadvantage of these types of meters is the time delay in reaching equilibrium and the tendency to hysteresis.

Today we find many brands of water activity meters in the market. Most of these meters are based on the relationship between ERH and the food system, but differ in their internal components and configuration of software used. One of the water activity meters most used today is the AcquaLab Series 3 Model TE, developed by Decagon Devices, which is based on the chilled-mirror dew point method. This instrument is a temperature controlled water activity meter that allows placement of the sample in a temperature stable environment without the use of an external water bath. The temperature can be selected on the screen and is monitored and controlled with thermoelectric components. Most of the older generations of water activity instruments are based on a temperature-controlled environment. Therefore, a margin of error greater than5% can be expected due to temperature variations. This equipment is highly recommended for measuring water activity in fruits and vegetables since it measures a wide range of water activity.

The major advantages of the chilled-mirror dew point method are accuracy, speed, ease of use and precision. The AquaLab's range is from 0.030 to $1.000a_w$, with a resolution of $\pm0.001a_w$ and accuracy of $\pm0.003a_w$. Measurement time is typically less than five minutes. Capacitance sensors have the advantage of being inexpensive, but are not usually as accurate or as fast as the chilled-mirror dew point method. Capacitive instruments measure over the entire water activity range 0 to 1.00 a_w, with a resolution of $\pm0.005a_w$ and accuracy of $\pm0.015a_w$. Some commercial instruments can complete measurements in five minutes while other electronic capacitive sensors usually require 30 to 90 minutes to reach equilibrium relative humidity conditions.

3.2 Intermediate Moisture Foods (IMF) concept

Traditional intermediate moisture foods (IMF) can be regarded as one of the oldest foods preserved by man. The mixing of ingredients to achieve a given a_w, that allowed safe storage while maintaining enough water for palatability, was only done, however, on an empirical basis. The work done by food scientists approximately three decades ago, in the search for convenient stable products through removal of water, resulted in the so-called modern intermediate moisture foods. These foods rely heavily on the addition of humectants and preservatives to prevent or reduce the growth of microorganisms. Since then, this category of products has been subjected to continuous revision and discussion.

Definitions of IMF in terms of a_w values and moisture content vary within wide limits (0.6-0.90 a_w, 10-50% moisture), and the addition of preservatives provides the margin of safety against spoilage organisms tolerant to low a_w. Of the food poisoning bacteria, *Staphylococcus aureus* is one of the organisms of high concern since it has been reported to tolerate a_w as low as 0.83-0.86 under aerobic conditions. Many of the considerations on the significance of microorganisms in IMF are made in terms of a_w limits for growth. However, microbial control in IMF does not only depend on a_w but on pH, *E*h, *F* and *T* values preservatives, competitive microflora, etc., which also exert an important effect on colonizing flora.

3.2.1 Fruits preserved under IMF concept

The application of IMF technology has been very successful in preserving fruits and vegetables without refrigeration in most Latin American countries. For instance, the addition of high amounts of sugar to fruits during processing will create a protective layer against microbial contamination after the heat process. The sugar acts as a water activity depressor limiting the capability of bacteria to grow in food. As described in Figure 3.1, IMF foods are those with a_w in the range of 0.65 to 0.90 and moisture content between 15% and 40%. Food products formulated under this concept are stable at room temperature without thermal processing and can be generally eaten without rehydration. Some processed fruits and vegetables are considered IMF foods. These include cabbage, carrots, horseradish, potatoes, strawberries, etc.; their water activities at 30°C follow:

Foods	a_w
Cabbage	0.64
	0.75
Carrots	0.64
	0.75
Horseradish	0.75
Potatoes	0.75
	0.64
Strawberries	0.65
	0.75

Under these conditions, bacterial growth is inhibited but some moulds and yeast may grow at a_w greater than 0.70. In addition, chemical preservatives are generally used to inhibit the growth of moulds and yeasts in fruits and vegetables.

3.2.2 Advantages and disadvantages of IMF preservation

Advantages:

Intermediate moisture foods have an a_w range of 0.65-0.90, and thus water activity is their primary hurdle to achieving microbial stability and safety. IMF foods are easy to prepare and store without refrigeration. They are energy efficient and relatively cheap. They are not readily subject to spoilage, even if packages have been damaged prior to opening, as with thermostabilized foods, because of low a_w. This is a plus for many developing countries, especially those in tropical climates with inadequate infrastructure for processing and storage, and offers marketing advantages for consumers all over the world.

Disadvantages:
Some IMF foods contain high levels of additives (i.e., nitrites sulphites, humectants, etc.) that may cause health concerns and possible legal problems. High sugar content is also a concern because of the high calorific intake. Therefore, efforts are been made to improve the quality of such foods by decreasing sugar and salt addition, as well as by increasing the moisture content and a_w, but without sacrificing the microbial stability and safety of products if stored without refrigeration. This may be achieved by an intelligent application of hurdles (Leistner, 1994).

Fruit products from intermediate moisture foods (IMF) appear to have potential markets. However, application of this technology to produce stable products at ambient temperature is limited by the high concentration of solutes required to reduce water activities to safe levels. This usually affects the sensory properties of the food.

3.3 Combined methods for preservation of fruits and vegetables

3.3.1 Why combined methods?

Food preserved by combined methods (hurdles) remains stable and safe even without refrigeration, and is high in sensory and nutritive value due to the gentle process applied. Hurdle technology is the term often applied when foods are preserved by a combination of processes. The hurdle includes temperature, water activity, redox potential, modified atmosphere, preservatives, etc. The concept is that for a given food the bacteria should not be able to "jump over" all of the hurdles present, and so should be inhibited. If several hurdles are used simultaneously, a gentle preservation could be applied, which nevertheless secures stable and safe foods of high sensory and nutritional properties. This is because different hurdles in a food often have a synergistic (enhancing) or additive effect. For instance, modified foods may be designed to require no refrigeration and thus save energy. On the other hand, preservatives (e.g., nitrite in meats) could be partially replaced by certain hurdles (such as water activity) in a food. Moreover, a hurdle could be used without affecting the integrity of food pieces (e.g., fruits) or in the application of high pressure for the preservation of other foods (e.g., juices). Hurdle technology is applicable both in large and small food industries. In general, hurdle technology is now widely used for food design in making new products according to the needs of processors and consumers. For instance, if energy preservation is the goal, then energy consumption hurdles such as refrigeration can be replaced by hurdles (a_w, pH, or E_h) that do not require energy and still ensure a stable and safe product.

The hurdle effect is an illustration of the fact that in most foods several factors (hurdles) contribute to stability and safety (Leistner, 1992). This hurdle effect is of fundamental importance for the preservation of food, since the hurdles in a stable product control microbial spoilage and food poisoning as well as undesirable fermentation.

3.3.2 General description of combined methods for fruits and vegetables

Increasing consumer demand for fresh quality products is turning processors to the so-called minimally processed products (MP), an attempt to combine freshness with convenience to the point that even the traditional whole, fresh fruit or vegetable is being packaged and marketed in ways formerly reserved for processed products (Tapia et al., 1996). According to these authors, the widely accepted concept of MP refrigerated fruits involves the idea of living respiring tissues. Because MP refrigerated products can be raw, the cells of the vegetative tissue may be alive and respiring (as in fruits and vegetables), and biochemical reactions can take place that lead to rapid senescence and/or quality changes. In these products, the primary spoilage mechanisms are microbial growth and physiological and biochemical changes, and in most cases, minimally processed foods are more perishable than the unprocessed raw materials from which they are made.

The technology for shelf–stable high moisture fruit products (HMFP) is based on a combination of inhibiting factors to combat the deleterious effects of microorganisms in fruits, including additional factors to reduce major quality losses from reactions. In order to select a combination of factors and levels, the type of microorganism and quality loss from reactions that might occur must be anticipated (Tapia et al., 1996). Minimal processing may encompass pre-cut refrigerated fruits, peeled refrigerated whole fruits, sous vide dishes, which may include pre-heated vegetables and fruits, cloudy and clarified refrigerated juices, freshly squeezed juices, etc. All of these products have special packaging requirements coupled with refrigeration (Tapia et al., 1996). These products, apart from special handling, preparation, and size reduction operations, might also require special distribution and utilization operations such as Controlled atmosphere/Modified atmosphere/air flow rate/vacuum storage (O_2, CO_2, N_2, CO, C_2H_2, H_2O controls), computer controlled warehousing, retailing and food service, communications network, etc. HMFP fruits are less sophisticated than MPR fruits and should be priced lower when introduced commercially (Tapia et al., 1996). Careful selection of these processes should of course be made to find the appropriate methods suited to a particular rural or village situation.

An example of the hurdle technology concept is presented in Figure 3.4, in which a comparison of HMFP, IMF and MPR fruits in terms of hurdle(s) involved is made. Example A represents an intermediate moisture fruit product containing two hurdles (pH, and a_W). The microorganisms cannot overcome (jump over) these hurdles, thus the food is microbiological stable. In this case, a_W is the most relevant hurdle exerting the strongest pressure against microbial proliferation of IMF. In the preservation system of HMFP (example B), it is obvious that a_W does not represent the hurdle of highest relevance against microbial proliferation; pH is the hurdle exerting the strongest selective pressure on microflora. As in example A, HMFP does not require refrigerated storage. In example C, the mild heat treatment T(t) is applied and the chemical preservative, P, added affects the growth and survival of the flora. With these considerations in mind, it is possible to understand and anticipate the types of microorganisms that could survive, as well as their behaviour and control in such fruits.

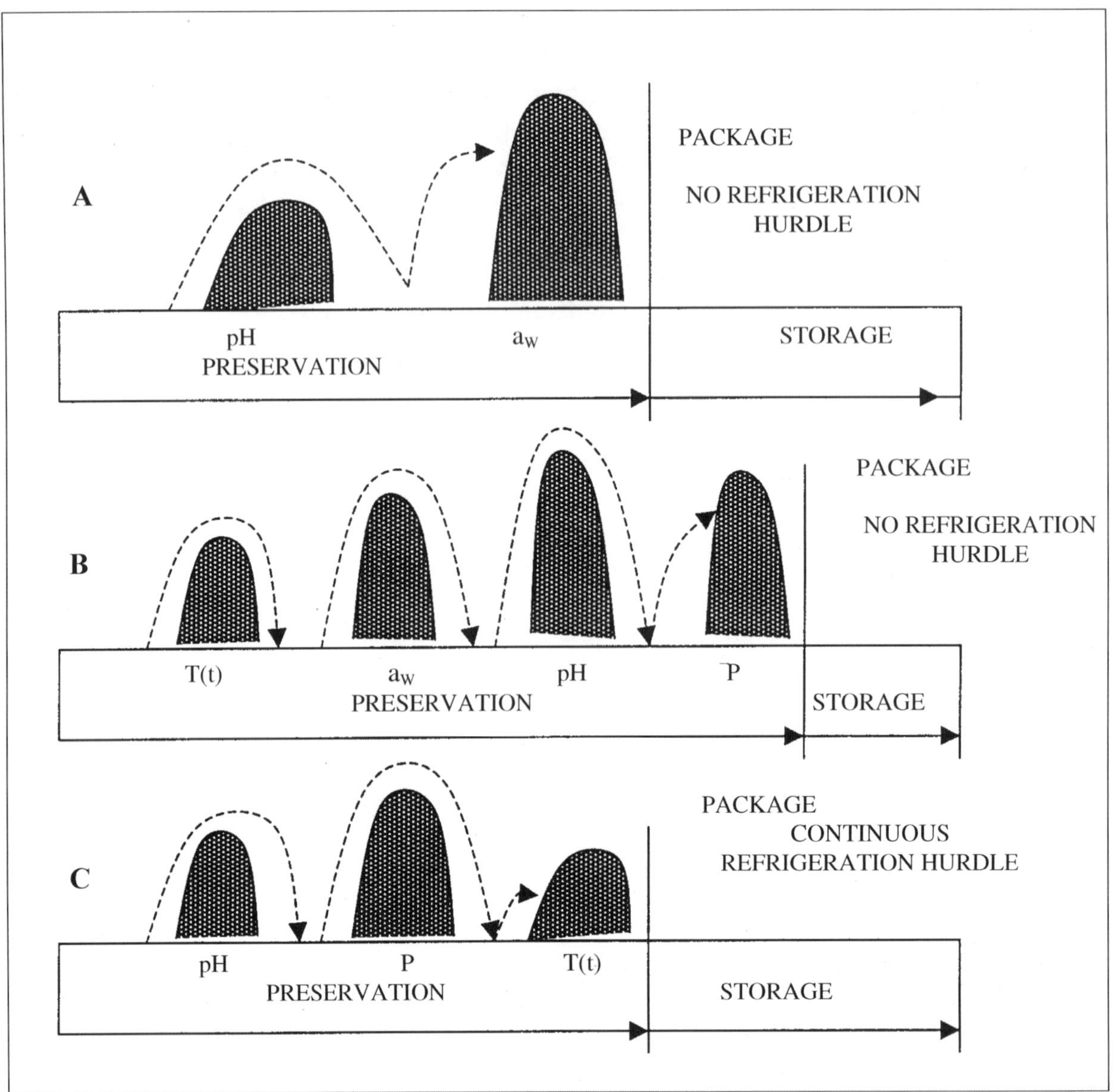

Figure 3.4. Schematic representation of hurdles: water activity (a_W), pH, preservatives (P), and slight heat treatment, T(t), involved in three fruit preservation systems. (A) an intermediate moisture fruit product; (B) a high-moisture fruit product; (C) a minimally processed refrigerated fruit product. (Adapted from Tapia et al., 1996)

3.3.3 Recommended substances to reduce a_w in fruits

3.3.3.1 Glucose

Glucose is not a very good humectant due to the lower water holding capacity (WHC), which makes it difficult to obtain the isotherm curve at low a_w.

3.3.3.2 Fructose

Fructose has a higher water activity reduction capacity and therefore is more desirable as a humectant in stabilizing food products.

3.3.3.3 Sucrose

Sucrose is one of the most studied sugars and is widely used in food systems, in the confectionary industry, both in the U.S. and Europe, but has a lower water activity reduction capacity compared to fructose.

The water reduction capacity of sugar and salts in their amorphous and anhydrous state at different a_W is presented in Table 3.1.

Table 3.1. Water activity reduction capacities of sugars and salts.

Moisture content (g H_2O/100 g Solids)								
	Anhydrous				Amorphous			
Sugars	a_W = 0.60	0.70	0.80	0.90	0.60	0.70	0.80	0.90
Sucrose	3.0	5.0	10.0	-	14.0	20.0	35.0	65.0
Glucose	1.0	3.5	7.5	12.5	1.0	3.5	8.0	22.0
Fructose	14.0	22.0	34.0	47.0	18.0	30.0	44.0	80.0
Lactose	0.01	0.01	0.05	0.10	4.5	4.7	4.7	-
Sorbitol (adsorption)	17.0	22.0	37.0	76.0	25.0	35.0	55.0	110.0
Corn syrup	-	-	-	-	14.0	20.0	30.0	54.0
Salts								
NaCl (adsorption)	0.1	0.1	130.0	585.0	-	-	-	-
NaCl (desorption)	-	-	385.0	590.0	-	-	-	-
KCl (adsorption)	0.1	0.1	0.1	0.1	-	-	-	-
KCl (desorption)	-	-	0.1	580.0	-	-	-	-

Source: Sloan and Labuza (1975).

3.3.3.4 Other humectants

Based solely on the water activity reduction capacity (Table 3.1), sorbitol and fructose are the most desirable humectants. Sucrose has the third best reduction capacity and lactose the poorest. The amorphous form absorbs more water at specific a_W than the corresponding crystalline form. As seen in Table 3.1, NaCl and KCl salts appear to be superior humectants at a high range of a_W. The increased a_W lowering ability exhibited by the salts may be explained by the smaller molecular weight, increasing the ability to bind or structure more water (Sloan and Labuza, 1975).

Other sugars used as humectants in food stability include lactose and sorbitol. The amorphous form absorbs more water at specific a_W than the crystalline form. Polyols are better humectants than sugars because of their greater water activity reduction capacity and are less hygroscopic than sugars. The most widely used polyols as humectants in foods are 1,3- butyleneglycol, propylene glycol, glycerol, and polyethylene glycol 400.

3.3.4 Recommended substances to reduce pH

3.3.4.1 Organic acids

Organic acids, whether naturally present in foods due to fermentation or intentionally added during processing, have been used for many years in food preservation. Some organic acids behave primarily as fungicides or fungistats, while others tend to be more effective at inhibiting bacterial growth. The mode of action of organic acids is related to the pH reduction of the substrate, acidification of internal components of cell membranes by ionization of the undissociated acid molecule, or disruption of substrate transport by alteration of cell membrane permeability. The undissociated portion of the acid molecule is primarily responsible for antimicrobial activity; therefore, effectiveness depends upon the dissociation constants (pKa) of the acid. Organic acids are generally more effective at low pH and high dissociation constants. The most commonly used organic acids in food preservation include: citric, succinic, malic, tartaric, benzoic, lactic, and propionic acids.

Citric acid is present in citrus fruits. It has been demonstrated that citric acid is more effective than acetic and lactic acids for inhibiting growth of thermophilic bacteria. Also, combinations of citric and ascorbic acids inhibit growth and toxin production *of C. botulinum* type B in vacuum-packed cooked potatoes.

Malic acid is widely found in fruits and vegetables. It inhibits the growth of yeasts and some bacteria due to a decrease in pH.

Tartaric acid is present in fruits such as grapes and pineapples. The antimicrobial activity of this acid is attributed to pH reduction.

Benzoic acid is the oldest and most commonly used preservative. It occurs naturally in cranberries, raspberries, plums, prunes, cinnamon, and cloves. As an additive, sodium salt in benzoic acid is suitable for foods and beverages with pH below 4.5. Benzoic acid is primarily used as an antifungal agent in fruit-based and fruit beverages, fruit products, bakery products, and margarine.

Lactic acid is not naturally present in foods; it is formed during fermentation of foods such as sauerkraut, pickles, olives, and some meats and cheeses by lactic acid bacteria. It has been reported that lactic acid inhibits the growth of spore forming bacteria at pH 5.0 but does not affect the growth of yeast and moulds.

Propionic acid occurs in foods by natural processing. It is found in Swiss cheese at concentrations up to 1%, produced by *Propionicbacterium shermanii.* The antimicrobial activity of propionic acid is primarily against moulds and bacteria.

3.3.4.2 Inorganic acids

Inorganic acids include hydrochloric, sulphuric, and phosphoric, the latter being the principal acid used in fruit and vegetable processing). They are mainly used as buffering agents, neutralizers, and cleaners.

3.3.4.3 Fermentation by-products

Fermentation by-products are formed during fermentation of fruits and vegetables, as in sauerkraut processing, pickling, and wine making. One by-product, lactic acid, is formed during fermentation of cabbage or cucumbers. This acid decreases the pH of fruits and vegetables, producing the characteristic flavour of sauerkraut, and acts as a controller of pathogens that may develop in the final fermented product.

3.3.5 Recommended chemicals to prevent browning

3.3.5.1 Sulphites, bisulphites, and metabisulphites

Sodium bisulphite is a potential browning inhibitor in fruit and vegetable products (e.g., peeled potatoes and apples). This preservative when used in food production can delay or prevent undesirable changes in the colour, flavour, and texture of fresh fruits and vegetables, potatoes, drinks, wine, etc. Potassium bisulphite is used in a similar way to sodium bisulphite, and is used in the food industry to prevent browning reactions in fruit and vegetable products.

Sulphites, bisulphites, and metabisulphites of both sodium and potassium together with gaseous sulphur dioxides are all chemically equivalent. Sulphite levels in processed foods are expressed as SO_2 equivalents, and range from zero to about 3000 ppm in dry weight. Dehydrated, light coloured fruits (e.g., apples, apricots, bleached raisins, pears, and peaches) contain the greatest amounts in this range. Dehydrated vegetables and prepared soup mixes range from a few hundred to about 2000 ppm; instant potatoes contain approximately 400 ppm. The dose for wine is about 100-400 ppm and for beer about 2-8 ppm. The maximum legal sulphite level in wines permitted by the Food and Drug Administration (FDA) is 300 ppm. In the U.S. most wines have a sulphite level of 100 ppm.

Sulphites are highly effective in controlling browning in fruits and vegetables, but are subject to regulatory restrictions because of adverse effects on health. Sulphites inhibit non-enzymatic browning by reacting with carbonyl intermediates, thereby preventing further reaction. Sulphite levels in foods vary widely depending on the application. Residual levels never exceed several hundred per million but could reach 100 ppm in some fruits and vegetables.

The maximum sulphur dioxide levels in fruit juices, dehydrated potatoes, and dried fruits permitted by the FDA are 300, 500, and 2000 ppm, respectively.

3.3.6 Recommended additives to inhibit microorganisms

3.3.6.1 Potassium sorbate

Potassium sorbate is a white crystalline powder that has greater solubility in water than sorbic acid, which may be used accordingly in making concentrates for dipping, spraying, or metering fruit and vegetable products. It has antimycotic actions similar to sorbic acid, but usually 25% more potassium sorbate must be used than sorbic acid to secure the same protection.

The common salt of potassium sorbate was developed because of its high solubility in water, which is 58.2% at 20°C (Sofos, 1989). In water, the salt hydrolysis yielded is the active form.

Stock solutions of potassium sorbate in water can be concentrated up to 50%, which can be mixed with liquid food products or diluted dips and sprays. Sorbates are effective in retarding the growth of many food spoilage organisms. Sorbates have many uses because of their milder taste, greater effectiveness, and broader pH range (up to 6.5), when compared to either benzoate or propianate. Thus, in foods with very low pH, sorbate levels as low as 200 ppm may give more than adequate protection. The solubility of potassium sorbate is 139 g/100 mL at 20°C; it can be applied in beverages, syrups, fruit juices, wines, jellies, jams, salads, pickles, etc.

3.3.6.2 Sodium benzoate

The use of sodium benzoate as a food preservative has been limited to products that are acid in nature. Therefore, it is mainly used as an antimycotic agent (most yeasts and moulds are inhibited by 0.05-0.1%). The benzoates and parabenzoates have been used primarily in fruit juices, chocolate syrup, candied fruit peel, pie fillings, pickled vegetables, relishes, horseradish, and cheeses. Sodium benzoate is more effective in food systems where the pH is as low as 4.0 or below.

3.3.6.3 Other additives

Other naturally antimicrobial compounds found in fruits and vegetables include:

Vanillin (4-hydroxi-3-methoxybenzaldehyde) is found primarily in vanilla beans and in the fruit of orchids *(Vanilla planifola*, *Vanilla pompona*, or *Vanilla tahitensis*). Vanillin is most active against moulds and non-lactic acid gram-positive bacteria. The effectiveness of vanillin against certain moulds such as *A. flavus*, *A. niger*, *A. ochraceus*, *or A. parasiticus* has been demonstrated in laboratory media, as well as its effectiveness against yeasts such as *Saccharomyces cerevisiae, Pichia membranaefaciens, Zygosaccharomyces bailii, Z. rouxii* and *Debaryomyces hansenii.*

Allicin is an antimicrobial present in the juice vapour of garlic. This compound is effective in inhibiting the growth of certain pathogenic bacteria such as *B. cereus*, C. botulinum, *E. coli*, *Salmonellae*, *Shigellae*, *S. aureus*, *A. flavus*, *Rhodotorula*, and *Saccharomyces*.

Cinnamon and eugenol are reported to have an inhibitory effect on the spores of *Bacillus anthracis*. Also, cinnamon was found to inhibit the growth of the aflatoxin of *A. parasiticus*. Aqueous clove infusions of 0.1 to 1.0% and 0.06% eugenol were reported to inhibit the growth of germinated spores of *B. subtilis* in nutrient agar.

Oregano, thyme, and rosemary have been found to have inhibitory activity against certain bacteria and moulds due to the presence of antimicrobial compounds in their essential oils (e.g., terpenes, carvacol, and thymol).

3.3.7 Recommended thermal treatment for food preservation

3.3.7.1 The role of heat

The main function of heat in food processing is to inactivate pathogenic and spoilage organisms, as well as enzyme inactivation to preserve foods and extend shelf life. Other advantages of heat processing include the destruction of anti-nutritional components of foods (e.g., trypsin inhibitors in legumes), improving the digestibility of proteins, gelatinization of starches, and the release of niacin. Higher temperatures for shorter periods achieved the same shelf life extension as food treated at lower temperatures and longer periods, and allowed retention of sensory and nutritional properties.

3.3.7.2 Hot water

Hot water plays an important role in the sanitation of food products before processing. Some food products are treated with hot water to eliminate insects, and to inactivate microorganisms and enzymes. Foods are retained in a water blancher at 70-100°C for a specific time and then removed to a dewatering and cooling system.

3.3.7.3 Steam

Steam is a more effective means than hot water for blanching foods such as fruits and vegetables. This method is especially suitable for foods with large areas of cut surfaces. It retains more soluble compounds and requires smaller volumes of waste for disposal than those from water blanchers. This is particularly so if air-, rather than water-cooling is used. Furthermore, steam blanchers are easier to clean and sterilize.

3.3.7.4 Effects of heat on aerobic and anaerobic mesophylic bacteria, yeasts, and moulds

Temperatures ranging from 10 to 15°C above the optimum temperature for growth will destroy vegetative cells of bacteria, yeasts, and moulds. Most vegetative cells, as well as viruses, are destroyed when subjected to temperatures of 60 to 80°C for an appropriate time. Somewhat higher temperatures may be needed for thermophilic or thermoduric microorganisms. All vegetative cells are killed in 10 min at 100°C and many spores are destroyed in 30 min at 100°C. Some spores, however, will resist heating at 100°C for several hours.

CHAPTER 4

EXTENSION OF THE INTERMEDIATE MOISTURE CONCEPT TO HIGH MOISTURE PRODUCTS

The importance of considering the combined action of decreased water activity with other preservation factors as a way to develop new improved foodstuffs has been studied. Leistner (1994) introduced the hurdle concept, or hurdle effect as discussed in the previous chapter, to illustrate the fact that in most foods, a combination of preservation parameters (hurdles) accounts for their final microbial stability and safety. Since then, these concepts have been improved to the point that depending on the acting hurdles of high relevance to a particular product, shelf-stability can be accomplished by a careful handling of complementary hurdles. For instance, the pH of IMF should be as low as palatability permits, and whenever possible, below pH 5.0. Undoubtedly, this imposes a limitation not only on colonizing microflora, but also on foodstuffs, since pH cannot be reduced in many products without flavour impairment. Even at low pH values and low a_w, certain yeast and mould species that can tolerate high solute concentrations might pose a risk to the stability of IMF.

Fruits are a good example of foodstuffs that accept pH reduction without affecting the flavour significantly. Important developments on IMF based on fruits and vegetables are reported elsewhere. The extensive research conducted in India by Dr. Jayaraman and co-workers has generated important information on this product category. Technological problems have prevented IMF from further development. Also, consumer health concerns associated with the high levels of humectants and preservatives used, have contributed to this situation. This last issue has become more important in recent years due to greater public awareness of food safety concerns. Additionally, consumers are searching for fresh-like characteristics in products. The food industry has responded to these demands with the so-called minimally processed fruits and vegetables, which have become a widespread industry. Consequently, safety considerations are being addressed seriously by food microbiologists.

Different approaches can be explored for obtaining shelf-stability and fresh-likeness in fruit products. Commercial, minimally processed fruits are fresh (with high moisture), and are prepared for convenient consumption and distribution to the consumer in a fresh-like state. Minimum processing includes preparation procedures such as washing, peeling, cutting, packing, etc., after which the fruit product is usually placed in refrigerated storage where its stability varies depending on the type of product, processing, and storage conditions. However, product stability without refrigeration is an important issue not only in developing countries but in industrialized countries as well. The principle used by Leistner for shelf-stable high moisture meats (a_w >0.90), where only mild heat treatment is used and the product still exhibits a long shelf life without refrigeration, can be applied to other foodstuffs. Fruits would be a good choice. Leistner states that for industrialized countries, production of shelf-stable products (SSP) is more attractive than IMF because the required a_w for SSP is not as low and less humectants and/or less drying of the product is necessary.

If fresh-like fruit is the goal, dehydration should not be used in processing. Reduction of a_w by addition of humectants should be employed at a minimum level to maintain the product in a high moisture state. To compensate for the high moisture left in the product (in terms of stability), a controlled blanching can be applied without affecting the sensory and nutritional properties; pH reductions can be made that will not impair flavour; and preservatives can be added to alleviate the risk of spoilage by microflora. In conjunction with the above mentioned factors, a slight thermal treatment, pH reduction, slight a_w reduction and the addition of antimicrobials (sorbic or benzoic acid, sulphite), all placed in context with the hurdle principle applied to fruits, make up an interesting alternative to IMF preservation of fruits, as well as to commercial minimally processed fruits.

Alzamora et al. (1995) conducted pioneer work aimed at obtaining shelf-stable peaches and pineapple. Considerable research has been made within the CYTED Program and the Multinational Project on Biotechnology and Food of the Organization of American States (OAS) in the area of combined methods geared to the development of shelf-stable high moisture fruit products.

Over the last decade, use of this approach has led to important developments of innovative technologies for obtaining shelf-stable "high moisture fruit products" (HMFP) storable for 3-8 months without refrigeration. These new technologies are based on a combination of inhibiting factors to combat the deleterious effects of microorganisms in fruits, including additional factors to diminish major quality loss in reactions rates. Slight reduction of water activity (a_w 0.94-0.98), control of pH (pH 3.0-4.1), mild heat treatment, addition of preservatives (concentrations $\leq$ 1,500 ppm), and antibrowning additives were the factors selected to formulate the preservation procedure. These techniques were preceded by the pioneer work of Leistner (1994) on the combined effects of several factors applied to meat products - named "hurdle" technology.

Microbiological preservation with these combined techniques, by gently applying individual stress factors to control microbial growth, avoid the severity of techniques based on the employment of only one conservation factor.

4.1 Preliminary operations

Preliminary operations involve washing, selecting, peeling, slicing, and general blanching of fresh fruits. Fresh produce must be processed between 4 and 48 hours after harvest to prevent the growth of spoilage microorganisms.

Washing: This operation involves eliminating dirt from the material before it passes through the processing line. Fruits are washed with potable water by immersion, spraying or brushing to eliminate the soil. Sodium hypochlorite is usually added to the water at a rate of 10% (v/v). The effectiveness of chlorine is enhanced by using a low pH, high temperature, pure water, and the correct contact time. A detailed description of this operation is given in Chapter 5.

Fruit selection: The cleaned product is selected for processing by separating the damaged fruits from those free of defects and disease. The fruit must be of a uniform size, form, colour, and maturity.

Peeling: This operation consists of removing the skin from the fruit (usually by hand) using a sharp knife. There are several peeling methods available, but on an industrial scale, peeling is normally accomplished mechanically (e.g., rotating carborundum drums) and chemically, or with high-pressure steam peelers. A detailed description of this operation is given in Chapter 5.

Slicing: This operation involves cutting the fruit into several uniform pieces, which is more convenient than handling the entire fruit. This is accomplished manually with a sharp knife or with special cutting machines that produce clean, neat slices.

Blanching: This is a critical control operation in the processing of high moisture fruit products (HMFP). It is an early step for processing of several fruits. Destruction of contaminating organisms is not the treatment's main objective, but it occurs nevertheless because the temperature used is lethal to yeast, most moulds, and aerobic natural flora. Many microorganisms can survive heat treatment but are sensitive to other hurdles like pH and water activity (a_w). A 60 to 99% reduction in the microbial load of HMFP for papaya, pineapple, strawberry, and mango has been reported. For mangoes, the microbial counts decreased from 14.3 x 10^3 cfu/g in the fresh fruit to 1.3 x 10^3 cfu/g after blanching. The blanching temperatures were between 85 and 100°C for very short periods, usually 3 to 5 minutes.

4.2 Desired a_w and syrup formulation

The desired a_w is determined by equilibrium of the components in the food system. This includes the addition of water, sugar (sucrose, glucose, or fructose), and chemicals such as citric acid, sodium bisulphite, and potassium sorbate, etc. The levels of sodium bisulphite and potassium sorbate in the system can be used at 150 and 1000 ppm, respectively. Once the system is in equilibrium, the a_w can be measured using an automatic water activity meter to an accuracy of + or – 0.005. These instruments are now available over specified ranges as laboratory or portable hand meters.

4.2.1 Calculus required

To determine the desired a_w in syrup ($a_{w\ equilibrium}$), the Ross equation is used:

$$a_{w\ equilibrium} = (a^{\circ}_{w})_{fruit} \bullet (a^{\circ}_{w})_{sugar} \qquad (1)$$

where $a^{\circ}_{w\ fruit}$ is the water activity of the fruit and $a^{\circ}_{w\ sugar}$ is the water activity of sugar, both calculated at the total molality of the system. The product of the molality of sucrose in the fruit water and solution must equal the desired water activity in equilibrium. The a°_{w} values of the sugar are obtained using the Norrish equation:

$$a^{\circ}_{w\ sucrose} = X_1 \exp\left(-kX_2^2\right) \qquad (2)$$

where k is a constant for sugars, X_1 and X_2 are the molar fractions of water and sugar, respectively. Some K values for common sugars and polyols are listed in Table 4.1.

Table 4.1. Values of Norrish constant for common sugars and polyols.

Sugars	***k***
Sucrose	6.47 ± 0.06
Maltose	4.54 ± 0.02
Glucose	2.25 ± 0.04
Lactose	10.2
Polyols	
Sorbitol	1.65 ± 0.14
Glycerol	1.16 ± 0.01
Mannitol	0.91 ± 0.27
Propylene Glycol	4.04
Arabitol	1.41

(From Barbosa-C novas and Vega -Mercado, 1996).

Phosphoric or citric acids are generally used to reduce the syrup's pH so that the final pH of the fruit-syrup system is in equilibrium in the desired range (3.0 to 4.1). Monitoring of a_W and pH in the fruit and syrup until constant values for these parameters are reached can determine the time to equilibrate the system. This may be from three to five days at constant room temperature depending on the size of fruit pieces.

Application of Norrish equation: example

The water activity of a sucrose-water solution (2.44:1 w/w) can be estimated by means of the Norrish equation. The mole fractions are: $X_1 = 0.887$ and $X_2 = 0.1125$. The Norrish constant (k) for sucrose is 6.47 (Table 4.1). Substituting X_1 and X_2 into the Norris equation results in the estimated water activity of the sucrose-water solution:

Approach

First, calculate the number of moles of water (MW = 18) and sucrose (MW = 342) and then determine the mole fraction for water and sucrose as described below. Insert values for X_1 and X_2 into the Norrish Equation to predict water activity of the sucrose-water solution.

X_1 = moles water/ (moles water + moles sucrose)

X_2 = moles sucrose/(moles sucrose + moles water)

Moles water = g water/Molecular weight water = 1/18 = 0.056

Moles sucrose = g sucrose/Molecular weight sucrose = 2.44/342 = 0.0071
Therefore,
$X_1 = 0.056/(0.056 + 0.0071) = 0.887$
$X_2 = 0.0071/(0.0071 + 0.056) = 0.1125$

Substituting these values into the Norrish equation, results in the estimated water activity of 0.817.

$$a^{\circ}_W\ sucrose = X_1 exp\left(-6.47X_2^2\right) = 0.887 \bullet \exp(-6.47 \times 0.1125^2) = 0.817$$

This value is within the range of IMF illustrated in Figure 4.1.

4.2.2 Water content vs. a_w relationship

Figures 4.1 and 4.2 represent typical curves that can be applied to most food systems for equilibrium water content (g water/g solid) versus water activity (% ERH). The graphs indicate the range in which foods can be adjusted. In general, dehydrated foods have less than 0.60 a_W; meanwhile, intermediate moisture foods (IMF) have water activity ranging between 0.62 and 0.92. Figure 4.1 shows that the water activity does not decrease much below 0.99 until the moisture content is reduced to 1 g H_20 per g of solid. A decrease in water activity or water content can be accomplished by drying, and by the addition of humectants, which reduces water activity through the effects of Raoult's law, or by the addition of dried ingredients such as starch, gums, or fibres, which interact with water through several mechanisms.

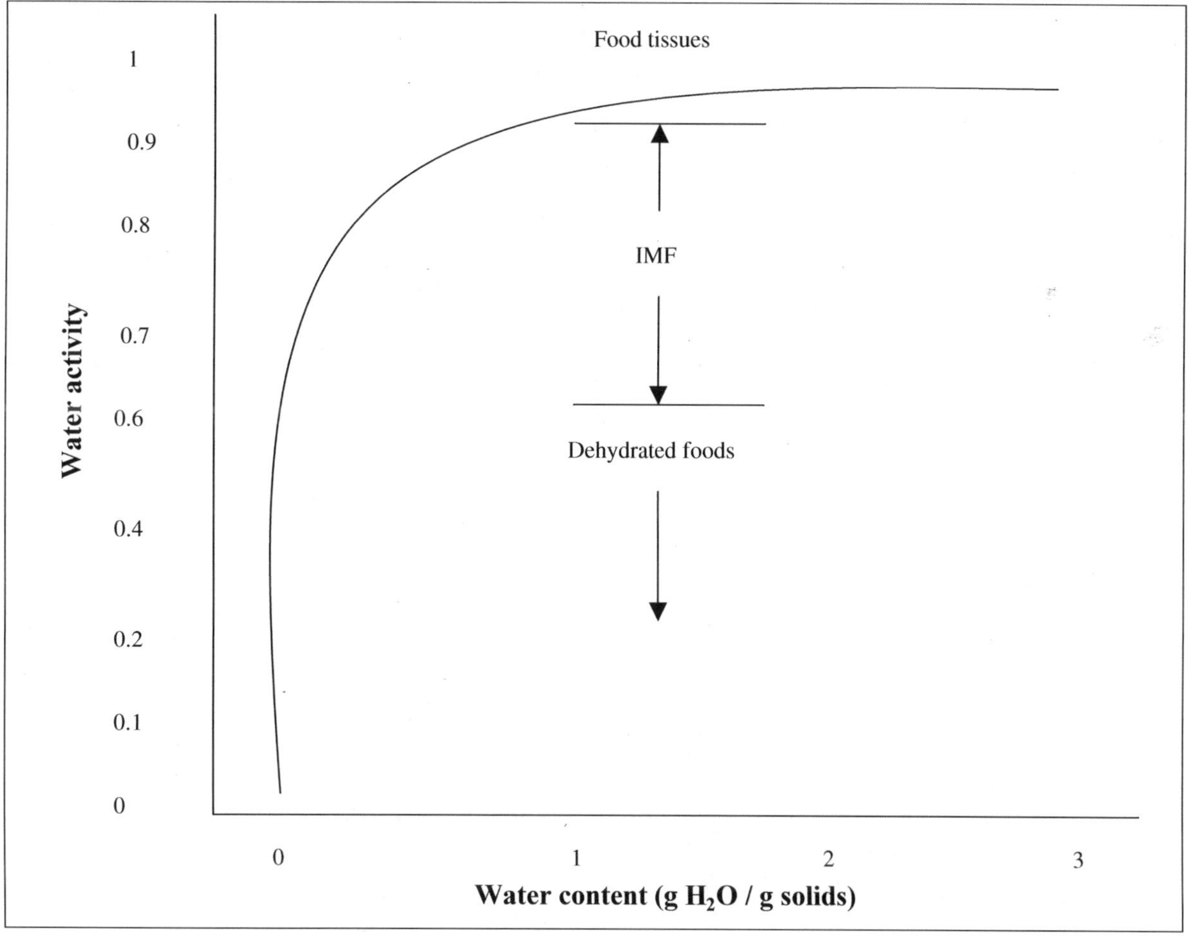

Figure 4.1 Typical equilibrium of water content vs. water activity in foods.

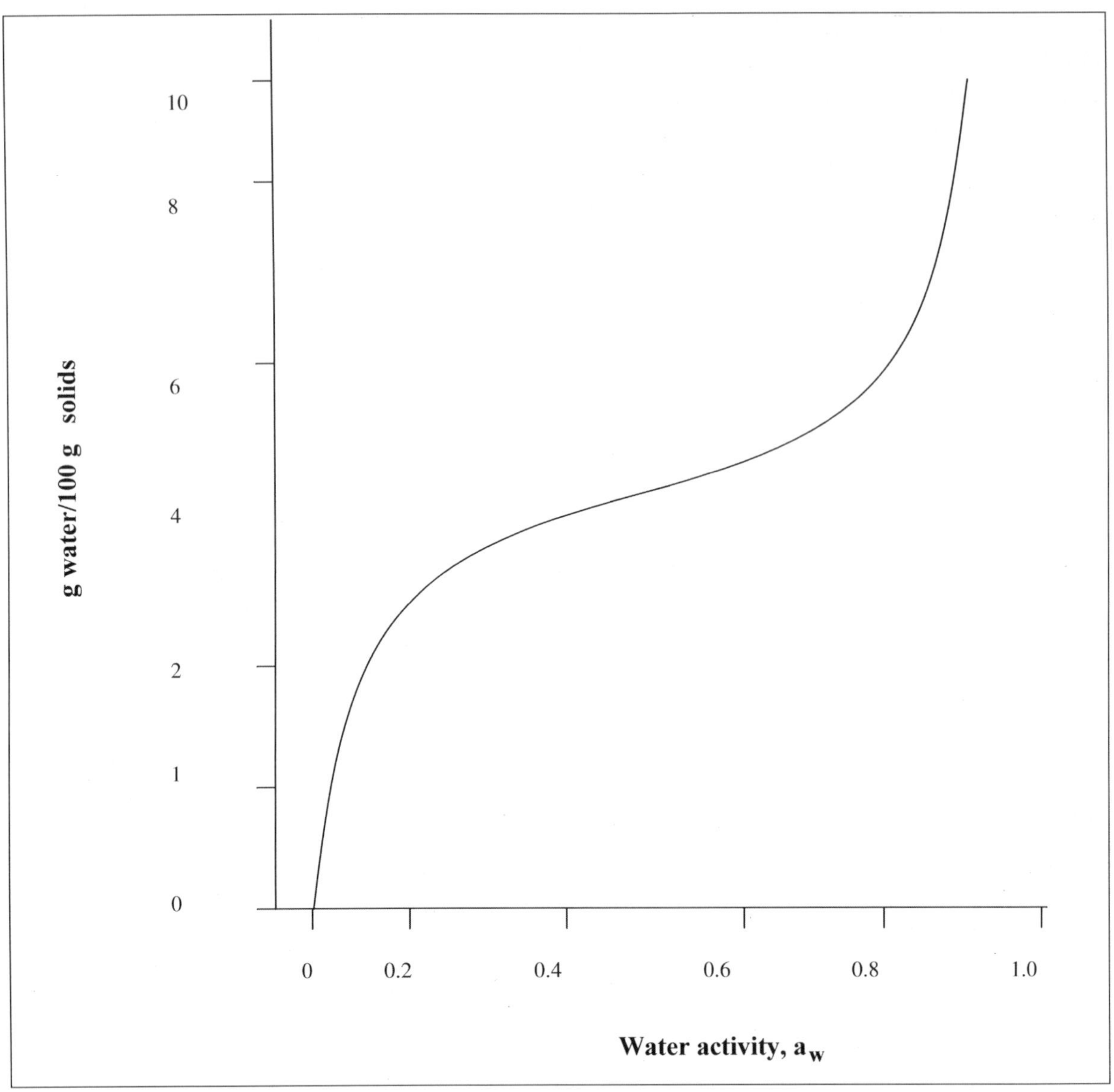

Figure 4.2 Equilibrium of water activity vs. moisture content, typical in foods. Lower region of isotherm.

4.3 Example of application

The general methodology will first be described before giving any specific examples of stabilized fruit and vegetable products by combined methods:

- Fruits and vegetables must be in a stage of unripeness.
- Only high quality fruits and vegetables are selected for processing.
- Non-edible matter is removed from the fruit: shells, leaves, tissues, stones, seeds.
- Raw material is thoroughly washed with potable water.
- Material is cut for final presentation into cubes, slices, etc.
- Pieces of fruits or vegetables are subjected to heat treatment by blanching, using saturated steam or boiling water for 1-2 minutes (depending on size of piece). The pieces are immediately cooled in water at 5-10ºC.

- After blanching and cooling, the fruit pieces are immediately drained and poured into a tank containing syrup or brine previously prepared. The fruit is immersed for 3-5 days until equilibrium is reached.
- The fruit pieces are drained and packed into glass or high density polyethylene plastic jars and covered with syrup. The product is now ready for marketing or direct consumer consumption.

Preparation of syrup or brine solution

To prepare the syrup or brine, a sufficient amount of sugar or salt is dissolved in water in order to reach the desired a_W. Concentrations of sulphur dioxide and potassium sorbate are prepared, reaching a final concentration of 100-150 ppm and 1000-1500 ppm, respectively. In the case of fruit products, citric or phosphoric acid are used to lower the pH of the syrup so that the final pH at equilibrium is in the range 3.0-4.1.

High moisture food products (HMFP) are very different from IMF products and need to be dehydrated. HMFP have a lower sugar concentration, 24-28% w/w compared to 20-40% w/w, and a higher moisture content, 55-75% w/w compared to 20-40% w/w, which makes them similar to canned food products. HMFP can be consumed directly after processing or bulk stored for processing out of season (Alzamora et al., 1995).

Several process flow-diagrams are given below for the preparation of HMFP (Figures 4.3 to 4.10). For each, the amount of sugar, salt, chemical preservatives (benzoates, sorbates, vanillin, etc.), browning agents (ascorbic acid, etc.), texturizers (calcium salts, etc.), etc., must be determined according to the weight of fruit used and the final levels required after equilibration of the product.

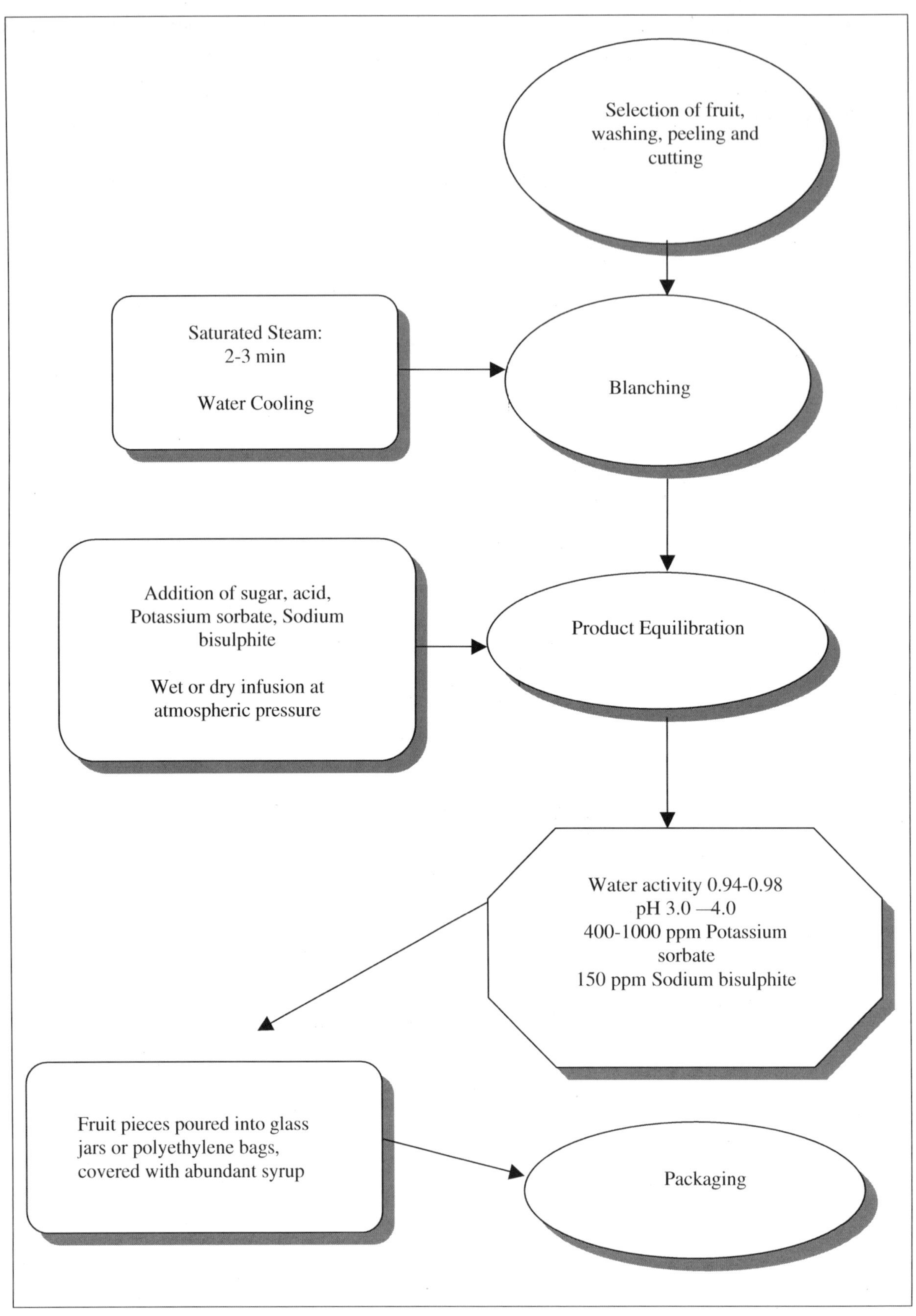

Figure 4.3 Preparation of shelf-stable HMFP (Welti et al, 2000).

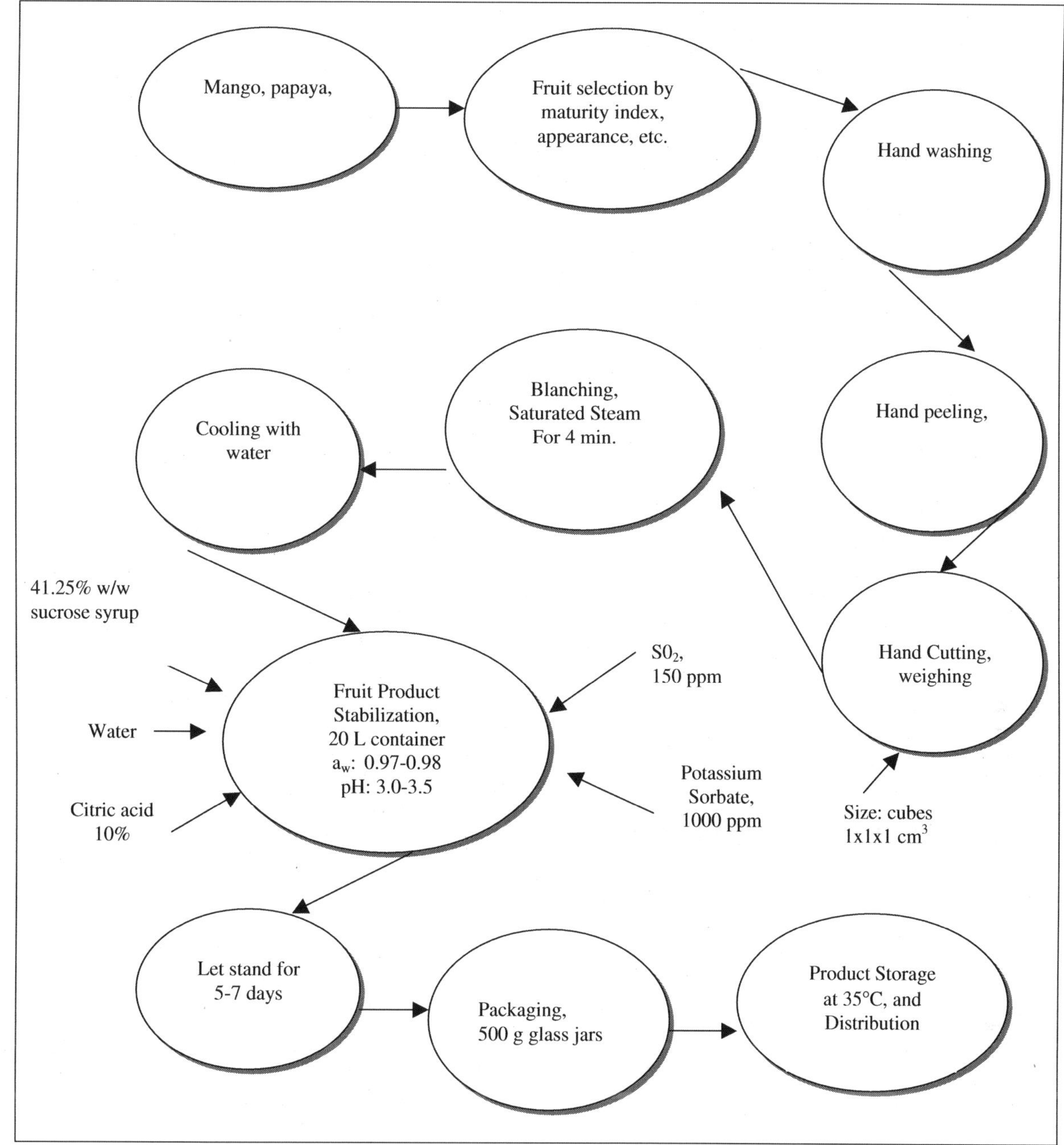

Figure 4.4 Schematic diagram for the preparation of shelf stable mango and papaya fruits by combined methods (Adapted from Diaz et al., 1993).

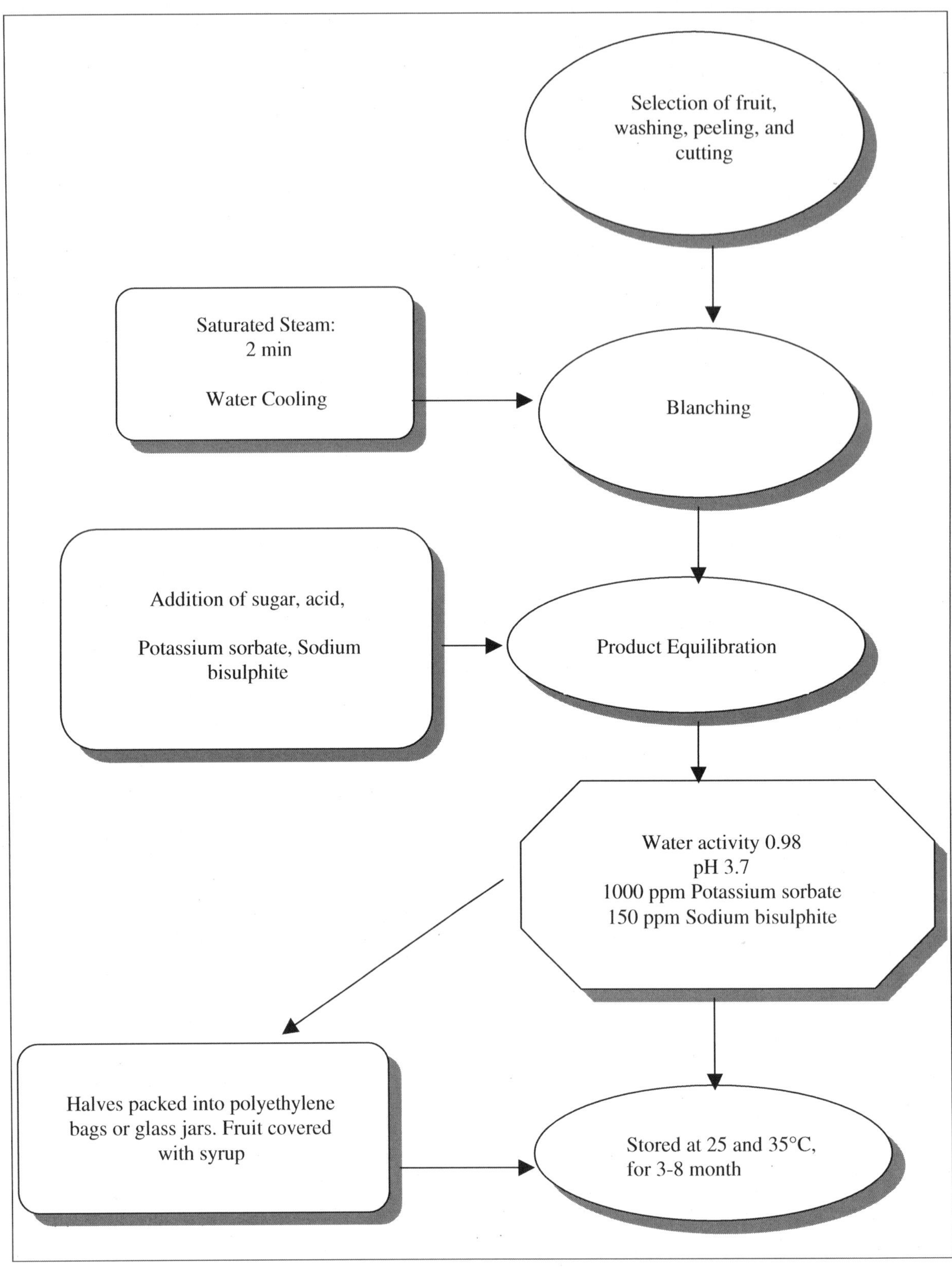

Figure 4.5 Flow process diagram for the preparation of shelf-stable high moisture peach halves (Welti et al., 2000).

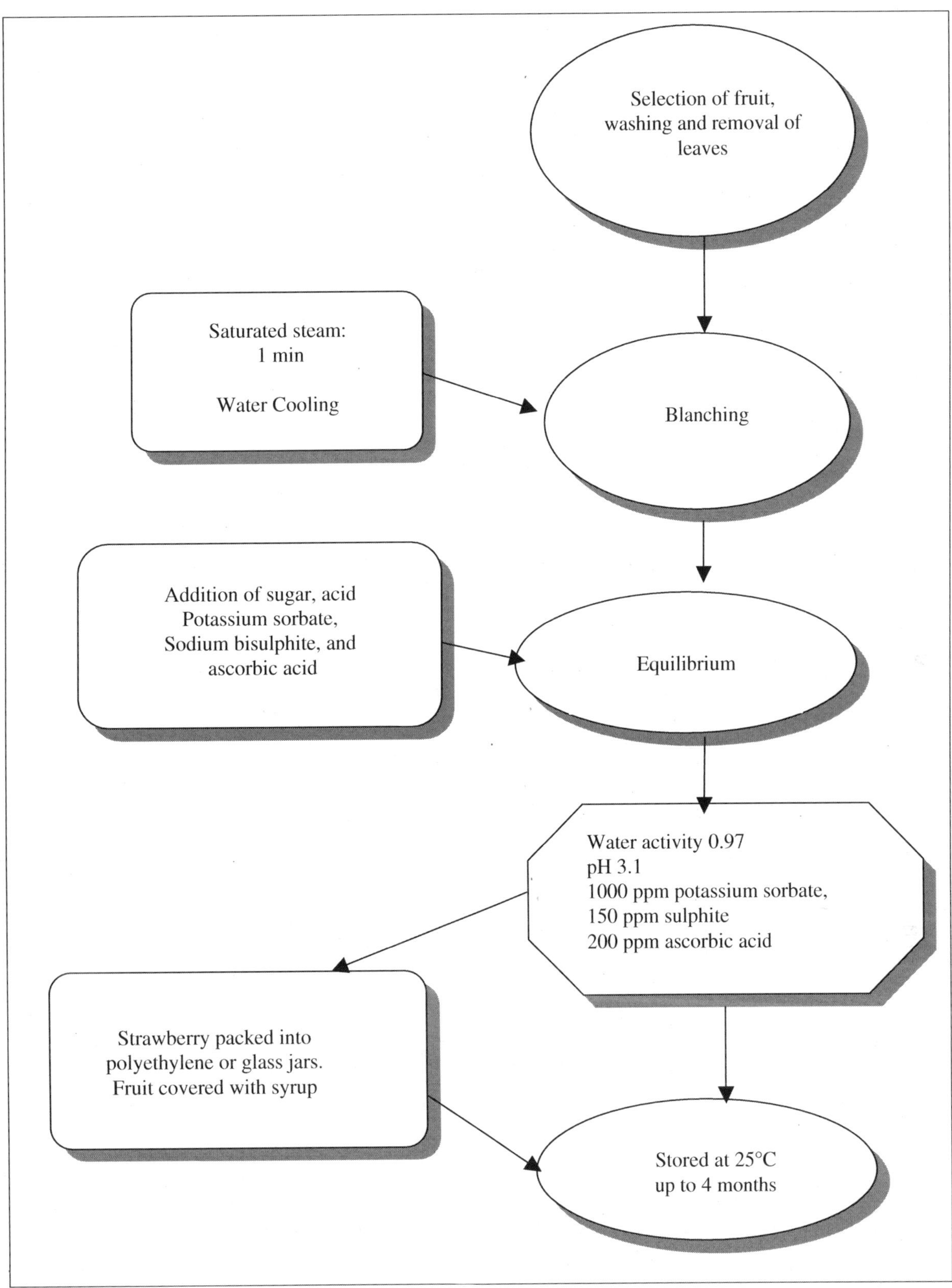

Figure 4.6 Flow process diagram for the preparation of shelf-stable high moisture whole strawberries (Welti et al., 2000).

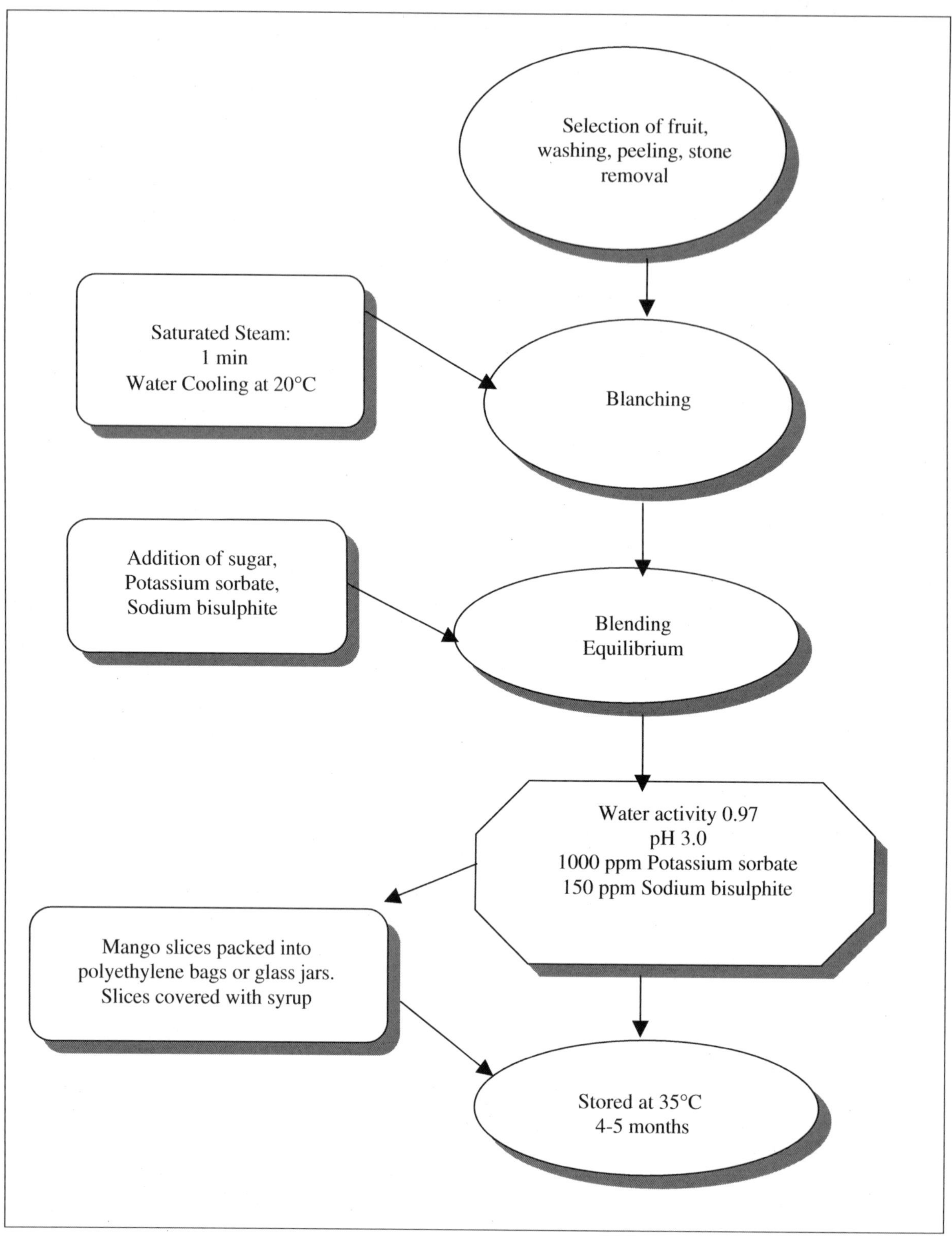

Figure 4.7 Flow process diagram for the preparation of stabilized high moisture mango slices (Welti et al., 2000).

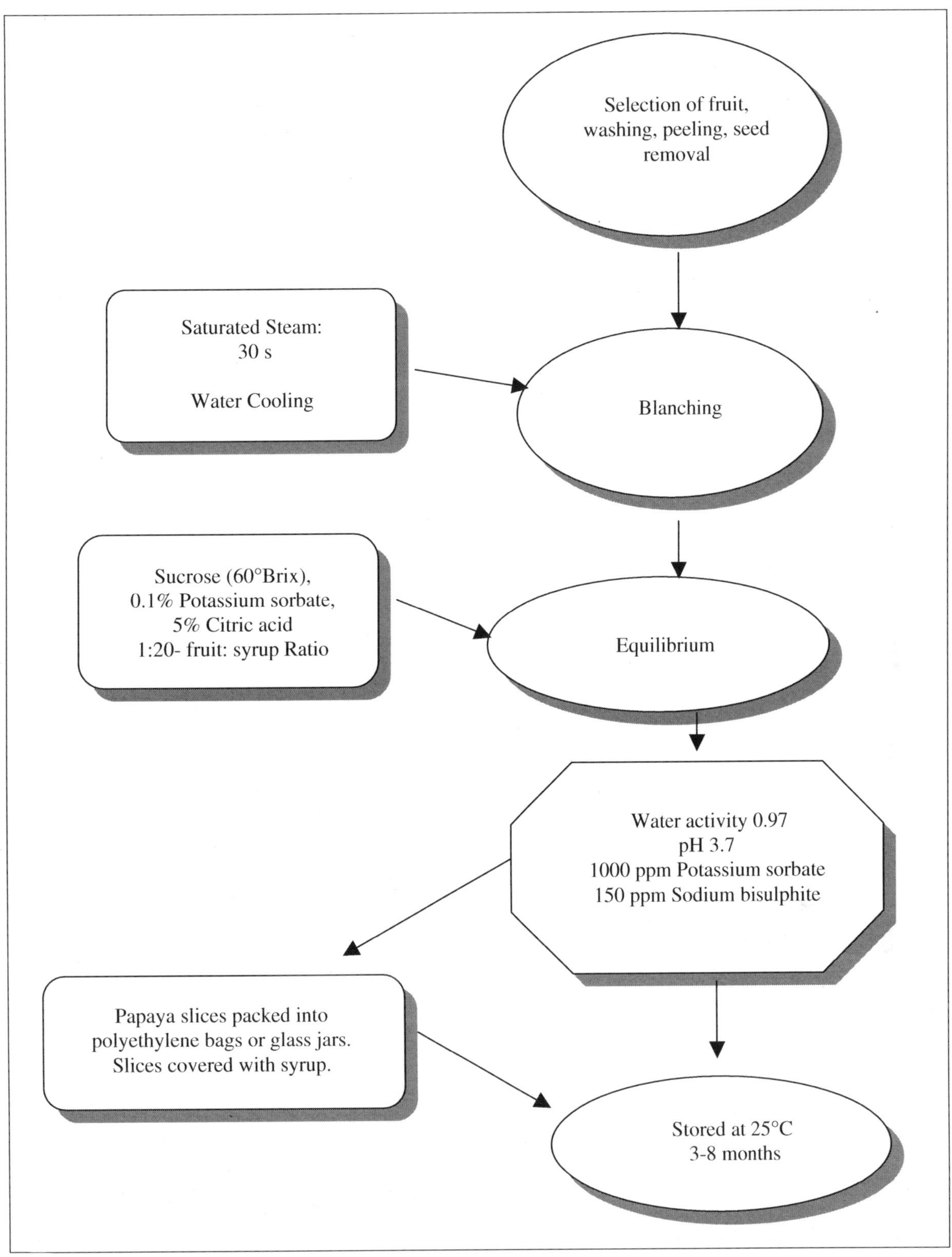

Figure 4.8 Flow process diagram for the preparation of stabilized high moisture papaya slices (Welti et al., 2000).

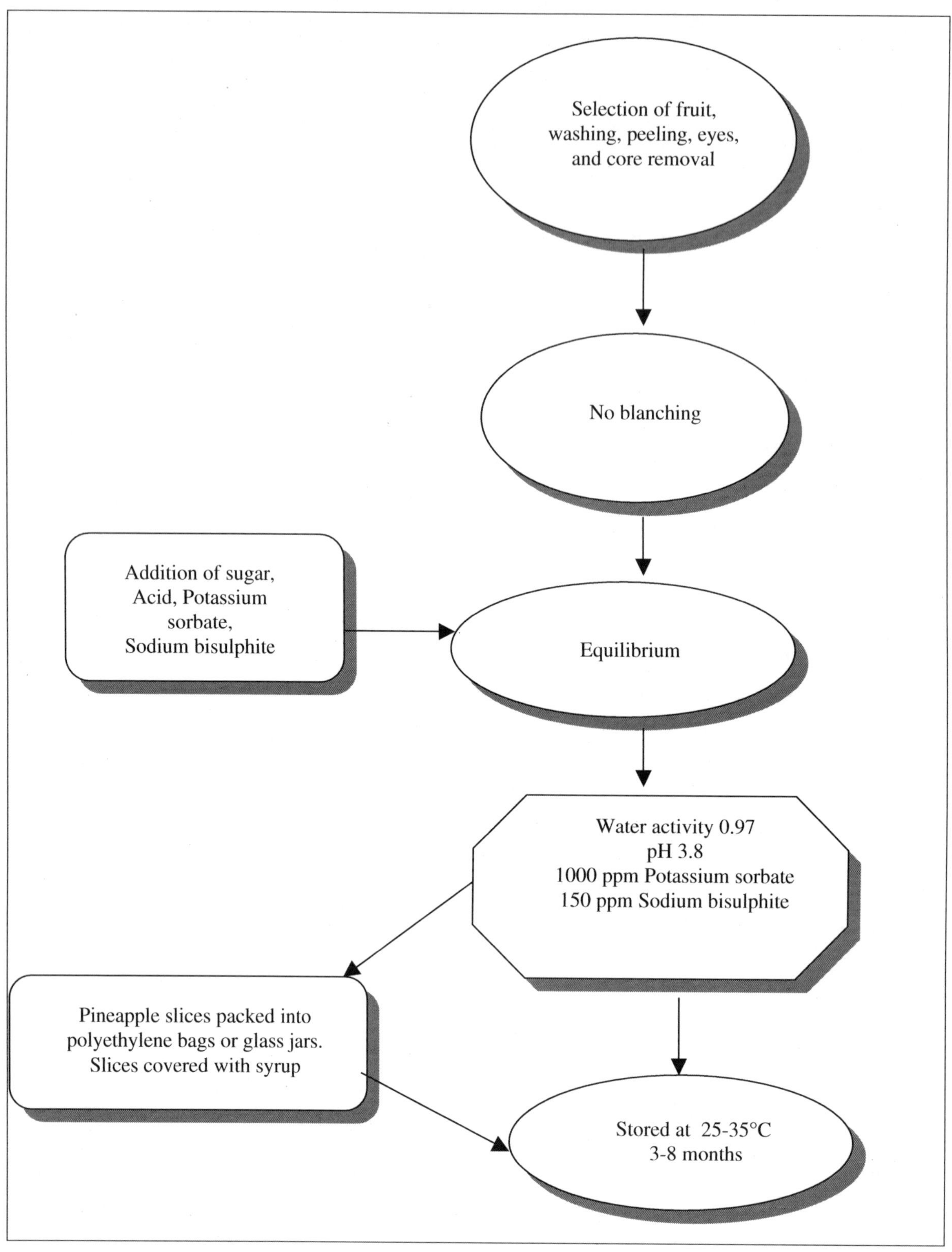

Figure 4.9 Flow process diagram for the preparation of shelf-stable high moisture pineapple slices (Welti et al., 2000).

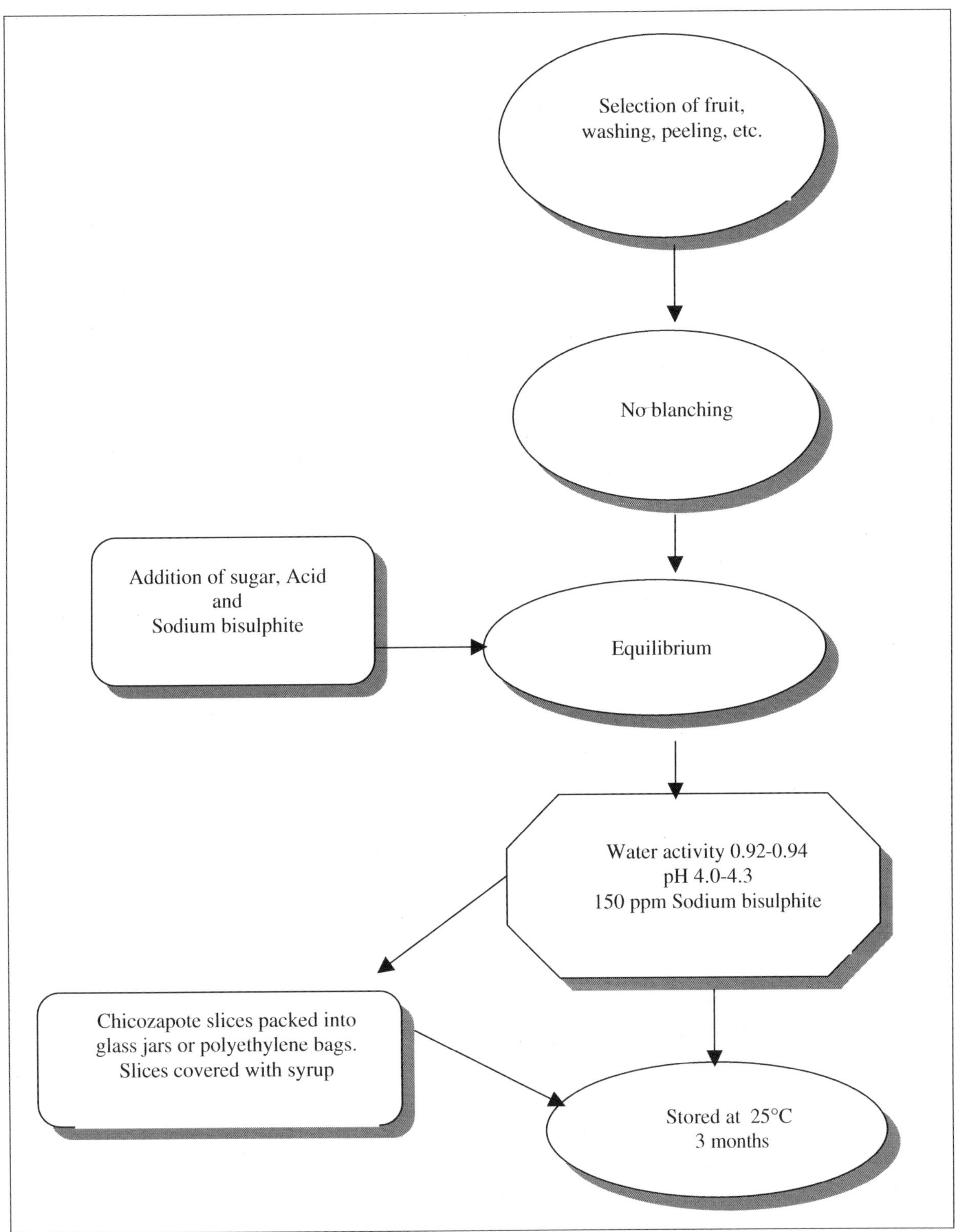

Figure 4.10 Flow process diagram for the preparation of shelf-stable high moisture chicozapote slices (Welti et al., 2000).

The high moisture fruit products stabilized by combined methods (Figure 4.4) were prepared from mango (*Mangifera indica L.*) var. "Bocado" and papaya (*Carica papaya L.*) var. "Criolla", grown in Venezuela. Mango and papaya fruits were cut into slices and chunks, subjected to steam blanching for 4 minutes, cooled in water, and stabilized in sucrose syrups

(42.25% w/w for mango and 33% w/w for papaya), with a fruit syrup ratio1:2 to attain equilibrium with a_W at 0.0.97 and 0.98, respectively. A final pH of 3.0 for mango and 3.5 for papaya was accomplished by adding citric acid. Sufficient sodium sulphite and potassium sorbate were also added to achieve equilibrium at 150 and 1000 ppm, respectively. The fruit products were equilibrated in 20 L plastic containers before packing into 500 g glass jars. The fruit products were stored for at least 30 days at 35°C, exhibiting good acceptability, microbial stability, and fresh-like appearance.

Sample calculation for preparation of a stable mango product:
Example 1: preserved mango pulp

The process conditions and ingredients required to prepare 20 kg of a stable mango product are: fruit pulp 16° Brix (16% soluble solids), acidity 0.5% (% citric acid). The fruit pulp is conditioned from 16° Brix (16% ss) to 40° Brix (40% ss) by adding sucrose. Sucrose is added to the pulp in order to act as a water activity depressor. The water activity of the pulp ranges from 0.97 to 0.98.

The mango is selected and processed as follows:
The fruit selected should be uniform in colour and size, firm and not bruised. Next it is washed with potable water, hand-peeled, and passed through a stainless steel pulp machine (5 mm mesh). The pulp is blanched at 80°C for 10 min. in a stainless steel kettle and cooled in running water. Afterwards, selected chemicals are added (sodium benzoate, 1000 ppm; sodium metabisulphite, 150 ppm), and acidity is adjusted with citric acid from 0.5% to 1% to obtain a product with a pH of approximately 3.6. (Barbosa-Cánovas et al., 1998; Tapia et al., 1996).

Calculation to obtain the amount of fruit pulp in the feed, sugar, citric acid, and free water in the final product:

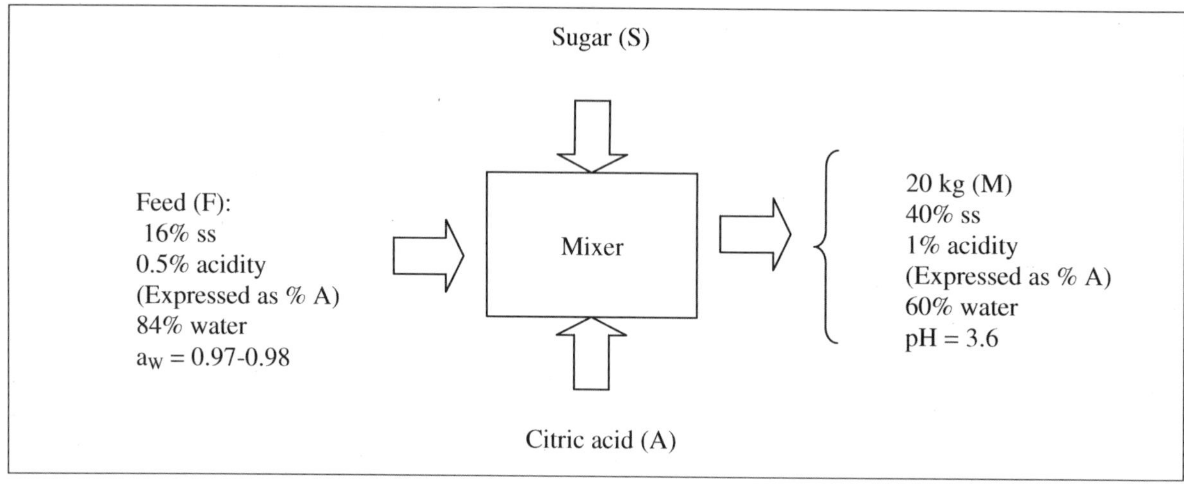

Solution:
Definition of terms:
F = kg of fruit pulp in the feed entering the mixer
S = kg of sugar (as sucrose) added to the fruit pulp
A = kg of citric acid added to the fruit pulp
W = kg of free water in the final product
M = kg of concentrated stabilized mango pulp

Overall Balance:

$F + S + A = M = 20$ (1)

Soluble Solids (ss) Balance:

$0.16 \cdot F + S = 20 \cdot 0.40 = 8.0$ (2)

Citric Acid Balance:

$0.005 \cdot F + A = 0.01 \cdot M = 0.01 \cdot 20 = 0.20$ (3)

Solving for S and A, from (2) and (3), and substituting into (1), we obtain:

$S = 8.0 - 0.16 \cdot F$

$A = 0.20 - 0.005 \cdot F$

$F + (8 - 0.16 \cdot F) + (0.2 - 0.005 \cdot F) = 20$

$0.835 \cdot \mathbf{F} = 20 - 8 - 0.2 = 11.8$

$F = \dfrac{11.8}{0.835}$ = 14.13 kg of fruit pulp in the feed

$S = 8.0 - (0.16 \cdot 14.13) = 5.74$ kg of sugar added to the pulp

$A = 0.20 - (0.005 \cdot 14.13) = 0.1294$ kg of citric added to the fruit pulp

Amount of water in the final product:

$W = 20 \cdot 0.60 = 12$ kg of free water in the stabilized mango pulp

The water activity of the mixture is predicted using Norrish Equation:

$$a_{wmixture} = X_1 \, Exp[-(K_2X_2^2 + K_3X_3^2)]$$

Where X_1 is the mole fraction of water, X_2 and X_3 are the mole fractions of sucrose and citric acid, respectively. K_2 and K_3 are constants for sucrose and citric acid. $K_2 = 6.47$ for sucrose and $K_3 = 6.20$ for citric acid (Barbosa-Cánovas et al., 1997).

Number of moles (n) = weight (g)/Molecular Weight

Moles of water (n_{Water}) = 12 /18 = 0.6667

Moles of Sucrose ($n_{Sucrose}$) = 5.74 /342 = 0.01678

Moles of Citric acid ($n_{Citric\ acid}$) = 0.1294 /192 = 0.00067395

Total number of moles (n_T) = $n_{Water} + n_{Sucrose} + n_{Citric\ acid}$

$n_T = 0.6667 + 0.01678 + 0.00067395 = 0.6842$

$$\text{Mole fraction (X)} = \frac{n}{n_{Total}}$$

$$X_1 = \frac{0.6667}{0.6842} = 0.9744$$

$$X_2 = \frac{0.01678}{0.6842} = 0.02453$$

$$X_3 = \frac{0.00067395}{0.6842} = 0.00098502$$

Substituting X_1, X_2, and X_3 into Norrish Equation we get the predicted water activity of the mixture as follow:

$$a_{wmixture} = 0.9744 \cdot Exp\,[-(6.47 \cdot (0.02453)^2) + (-6.20 \cdot (0.00098502)^2)]$$
$$= 0.9744 \cdot Exp\,[(-0.003893 - 0.000006015)]$$
$$= 0.9744 \cdot Exp\,[(-0.003899)] = 0.9744 \cdot [0.9961] = 0.97$$

The water activity at equilibrium between the fruit pulp and syrup is attained by application of the Ross Equation as follows:

$$A_{w\ equilibrium} = (a^{\circ}_{w})_{fruit} \bullet (a^{\circ}_{w})_{mixture} = (0.98).(0.97) = 0.95$$

Example 2: preserved pineapple slices

Figure 4.11 shows a flow chart for pineapple slices as an example of HMFP. Ripe pineapples are washed, cut into slices 2 cm thick, blanched in saturated steam for 2 min, cooled in water at 20°C, and immersed in glucose syrup. Sodium bisulphite and potassium sorbate are added to give 150 ppm and 1,000 ppm concentration, respectively. Glucose concentration in the syrup is calculated using the Ross equation (Barbosa-Cánovas and Vega-Mercado, 1996) to attain the a_w equilibrium value (0.97) between pineapple slices and syrup:

$$a_{w\ equilibrium} = (a^{\circ}_{w})_{pineapple} \bullet (a^{\circ}_{w})_{glucose} \quad (1)$$

where $(a_w)^{\circ}$ is the water activity of the fresh-fruit (≈ 0.99) and $(a_w)^{\circ}$glucose is the water activity of the sugar solution. Both water in the fruit and water in the solution are at the same molality.

Phosphoric acid is used to reduce the pH of the syrup to 2.76, with the final pH value for pineapple syrup at equilibrium 3.10. After equilibration (≈ 3 days for slices 2 cm thick), the fruit slices are drained, leaving only enough syrup to cover the product. The tanks containing the preserved fruits are held at constant room temperature during storage, resulting in a shelf life of at least 4 months.

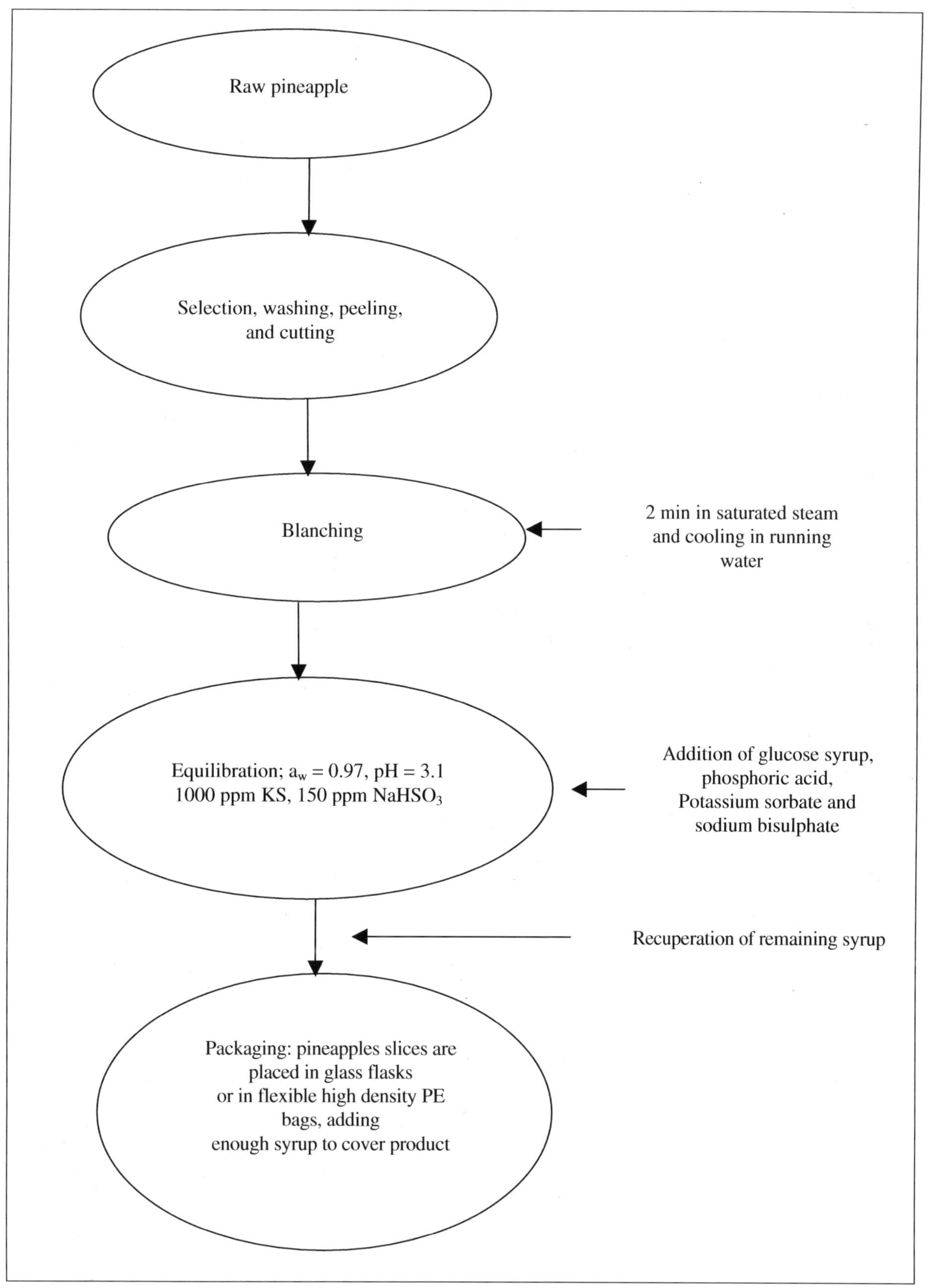

Figure 4.11 Flow diagram for the production of shelf-stable high moisture pineapple.

4.4 Packaging methods for minimally processed products

The purpose of food packaging is to maintain quality and to obtain shelf life extension of products by reducing mechanical damage and retarding microbial spoilage. Three types of packaging methods exist for minimally processed products: unit packaging, transport packaging, and loading packaging. Other packaging methods are vacuum and modified atmospheres.

4.4.1 Packaging with small units

This type of packaging method uses (1) closed plastic bags, (2) rigid or semi-rigid plastic trays zipped in upper part with polymeric plastic film, (3) covered trays for distribution of products to institutions (e.g., hotels, restaurants, and food shops) and small business consumer markets, (4) perforated or unperforated PE or PVC bags, (5) shallow trays, (6) cartons, and (7) thermo-formed plastic tubs or expanded PS containers covered/sealed with polymeric film (Wiley, 1997).

Two of the main requirements for this type of packaging are its permeability characteristics to any gases present and to water vapour. Other important considerations include: appearance (brightness and transparency), texture, resistance to water permeability, resistance to impact and deformation, thermo-seal capacity, ease in forming/fabricating/filling, and utilization of production equipment. Plastic containers are also light weight, sometimes reusable, tough, hygienic, and rigid containers can be stacked.

4.4.2 Transport the package

Packaging for transport of products is dominated by sealed cartons made from corrugated paper (Wiley, 1997). These types of packages provide good resistance to mechanical damage of fresh fruit, and facilitate manual handling of fresh fruits during transportation to markets. The cartons are made of paper, > 0.2 mm thick, obtained from vegetable cellulose bonded either in three layers with the middle one corrugated or in five layers with the second and fourth layers corrugated. Both systems provide a strong and rigid material.

4.4.3 Loading packaging units

This type of packaging implies the use of palletization of packages to reduce the cost of handling. In this way, the mechanical work of loading and unloading by carriers is facilitated, permitting better utilization of storage space and reducing mechanical damage during transportation.

4.4.4 Vacuum and modified atmosphere packaging

Vacuum packaging of fresh commodities involves eliminating (at least some) the air in the package using a suction machine. This method reduces the level of both oxygen and nitrogen in the package, prolonging the shelf life of fruits for extended periods.

Vacuum packaging is used in modified atmosphere packaging (MAP) of fruits and vegetables. The basic principle behind modified atmosphere packaging (MAP) is that a modified atmosphere can be created passively by correctly using permeable packaging materials, or actively, by using a special gas mixture combined with such materials. The purpose of both is to create an optimal gas balance inside the package, where the respiration activity of a product is as low as possible; on the other hand, the oxygen concentration and carbon dioxide levels are not detrimental to the product. In general, the aim is to have a gas composition of 2-5% CO_2, 2-5% O_2, and the rest nitrogen. A problem that arises when using MAP is the restricted availability of permeable material in the market, as only a few materials are permeable enough to match the respiration of fruits and vegetables. Most films do not result in optimal O_2 and CO_2 atmospheres, especially when the product has high respiration. However, one solution is to make microholes of defined sizes and defined quantity in the material to avoid anaerobiosis. Other solutions are to combine ethylene vinyl acetate with orientated polypropylene and low-density polyethylene, or to combine ceramic material with polyethylene. Both composite materials have significantly higher gas permeability than polyethylene or orientated polypropylene. They are used a lot in the packaging of salads, although gas permeability should be higher.

One interesting MAP method is called moderate vacuum packaging (MVP). In this system, respiring produce is packed into a rigid, airtight container at less than 0.4 of normal atmospheric pressure (40 kPa) and stored at refrigerated temperatures (4-7°C). The initial gas composition is that of normal air (21% O_2, 0.04 CO_2, and 78% N_2) but is at reduced partial gas pressure. The lower O_2 availability stabilizes the produce quality by slowing its metabolism and the growth of microorganisms.

4.5 Transport, storage, and use of fruits preserved by combined methods

4.5.1 Open vs. refrigerated vehicles

Open vehicles are mainly used to transport fresh produce over short distances from the field to packinghouses, retail markets, or the processing plant directly. The fruit must be protected against mechanical damage and sunlight. Therefore, the produce should be transported at night or in the early morning. Refrigerated vehicles should to be used to transport fruits. In this case, the vehicle must be equipped with an efficient cooling system, adequate distribution and circulation of air, relative humidity and temperature sensors, and it must be well insulated

4.5.2 Unloading

Unloading of fruits from vehicles can be done by hand or mechanical means. Forklifts are used to unload vehicles in which packages of fruits have been palletized. During unloading, care must be taken in handling the packages to avoid dropping, which can cause damage to the package and bruising of the fruit upon impact. Impact injury may not be visible on the surface; so careful control is needed to prevent its occurrence.

4.5.3 Storage temperature vs. shelf life

Refrigeration is the largest hurdle for MPF and the most difficult to control. During transport, handling, and storage of fruits by consumers, temperature is often not adequately maintained, resulting in spoilage. Food products exposed to elevated temperatures where refrigeration is the only factor of preservation are more susceptible to damage and spoilage, and thus the shelf life is very short.

Optimum refrigeration temperatures for fruits and vegetables vary widely. Some authors suggest between 10 and 15°C for cooling and between 2 and 5°C for refrigeration. Table 4.2 exhibits the optimum temperatures for storage of refrigerated fruits. Data is given for fresh products but the temperatures could change according to the process applied to a particular fruit. The data for recommended shelf life and the safety of minimally processed refrigerated fruits (MPRF) is still not available for public use. In general, MPRF products are classified as food products with prolonged shelf life where refrigeration is the preservation method most commonly used for this purpose.

The stability of fruits without refrigeration is an important issue in developing and industrialized countries. Minimally processed refrigerated fruits (MPRF) are not shelf stable at ambient temperatures and should be distributed and marketed in a reliable cold chain for safety and retention of sensory and nutritional quality. Hurdle technology has proved effective in preserving tropical and sub-tropical fruits with fresh-like properties. This technique includes blanching as an MP preservation method and excludes the use of refrigeration.

Table 4.2 Optimum temperatures for storage of refrigerated fruits.

FRUIT	TEMPERATURE
Apples	30-31°F (-1.1 to -0.6°C)
Varieties sensitive to refrigeration	38-40°F (3.3. to 4.4°C)
Apricots	31-32°C (-0.6 to 0°C)
Green plantains	56-58°F (13.3 to 14.4°C)
Berries:	31-32°F (-0.6 to 0°C)
Bush berries, blueberries, strawberries	32°C (0°C)
Cherries	30-32°F (1.1 to 0°C)
Citrus fruits:	
Grapefruits	58-60°F (14.4 to 15.6°C)
Lemons	58-60°F (14.4 to 15.6°C)
Limes	45-50°F (7.2 to 10°C)
Oranges	38-44°F (3.3 to 6.7°C)
Tangerines	32°F (0°C)
Coconuts	32-35°F (-0.6 to 0°C)
Dates	32°F (0°C)
Figs	31-32°F (-0.6 to 0°C)
Grapes	30-31°F (-1.1 to 0.6°C)
Mangoes	55°F (12.8°C)
Melons:	
Honeydew	45-50°F (7.2- to 10°C)
Cantaloupe	32-40°F (0- to 4.4°C)
Watermelon	40-32°F (-0.6 to 0°C)
Papayas	31°F-32°F (-0.6 to 0°C)
Peaches and nectarines	32°F (0°C)
Pears	29-31°F (-1.7 to -0.6°C)
Pineapples	45-47°F (7.6 to 8.3°C)
Prunes	31°F-32°F (-0.6 to 0°C)
Grenade	32°F (0°C)
Quince fruit	32°F (0°C)

Source: Wiley (1997).

Table 4.3 is a compilation of combined methods, and storage temperatures and shelf life, for minimally processed tropical fruits successfully developed in some Latin American countries, such as Argentina, Chile, Mexico, and Venezuela.

As can be seen in Table 4.3, the shelf life of high moisture fruits or purées is extended from at least 3 months to 8 months at room temperature. These fruit products are quite different from intermediate moisture fruits (high sugar candied fruits) because of a lower sugar concentration

(24-28% w/w vs. ≈ 70% w/w red. sugars) and higher moisture content (55-77% w/w vs. 20-40% w/w) that resembles canned fruit. They can be eaten as received or used as bulk for out-of-season processing, in confectionery, bakery goods, and dairy products, or for preserves, jams, and jellies. Fruit pieces can also be utilized as ingredients for salads, barbecues, pizzas, and fruit drink formulations.

Table 4.3 Combined methods for preserving tropical fruit with minimal processing.

FRUIT	**COMBINED METHOD**	**TEMPERATURE (°C)**	**SHELF LIFE (Month)**
Peach, Sliced, halves, or whole	Blanching (steam, 2 min.) a_w = 0.98 (sucrose) pH = 3.7 $NaHSO_3$ = 150 ppm KS = 1000 ppm	35	3
Peach, halves	Blanching (steam, 2 min.) a_w = 0.94 (glucose) pH = 3.5 $NaHSO_3$ = 150 ppm KS = 1000 ppm	20 or 30	4
Pineapple, sliced or whole	Blanching (steam, 2 min.) a_w = 0.97 (glucose) pH = 3.1 $NaHSO_3$ = 150 ppm KS = 1000 ppm	27	4
Mango	Blanching (steam, 4 min.) a_w = 0.97 (sucrose) pH = 3.0 $NaHSO_3$ = 150 ppm KS = 1000 ppm	35	4.5
Papaya	Blanching (steam, 30 sec.) a_w = 0.98 (sucrose) pH = 3.5 $NaHSO_3$ = 150 ppm KS = 1000 ppm	25	5
Strawberry	Blanching (steam, 1 min.) a_w = 0.97 (sucrose) or 0.95 (glucose) pH = 3.1 AA = 200 ppm $NaHSO_3$ = 150 ppm KS = 1000 ppm	25	4
Pomalaca	Blanching (85°C, 5min.) a_w = 0.97 (sucrose) pH = 3.5 SO_2 = 180 ppm KS = 1300 ppm Hot filling	35	3

FRUIT	COMBINED METHOD	TEMPERATURE (°C)	SHELF LIFE (Month)
Purees Banana	Blanching (steam, 1 min.) a_w= 0.97 (glucose) pH = 3.4 AA = 250 ppm $NaHSO_3$ = 400 ppm KS = 100 ppm Mild heat treatment (100°C, 1 min.)	27	3.5
Mango	Blanching (80°C, 10 min.) a_w = 0.985a pH = 3.6 SMB = 150 ppm SB = 1000 ppm	30-35	3
Papaya	Blanching (steam, 3 min.) a_w = 0.98 (sucrose) pH = 4.1 KS = 1000 ppm	35	4
Plum	Blanching (steam, 3 min.) a_w = 0.98 (sucrose) pH = 3.0 KS = 1000 ppm	25	4
Passion fruit	Blanching (steam, 3 min.) a_w = 0.98 (sucrose) pH = 3.0 SO_2 = 150 ppm KS = 400 ppm	30	4
Passion Fruit	a_w = 0.94 (sucrose) pH = 3.4 Heat treatment (85°C, 2 min.) $Na_2S_2O_3$ = 150 ppm KS = 1500 ppm Hot filling (60°C)	35	6
Tamarind	a_w = 0.96 (sucrose) pH = 2.5 Heat treatment (85°C, 2 min.) $Na_2S_2O_3$ = 150 ppm KS = 1500 ppm Hot filling (60°C)	35	6

KS = potassium sorbate; AA = ascorbic acid; SB = sodium benzoate; SMB = so dium metabisulphite. (From Tapia et al.,1996)

4.5.4 Repackaging considerations

Minimally processed fruit products can be repackaged from bulk containers into small packages such as glass or plastic jars, and high-density polyethylene bags for retail markets and consumer distribution. The stabilized fruit products can be processed in the form of slices, chunks, whole fruit, marmalades, or nectars.

4.5.5 Syrup reconstitution and utilization

Syrup reconstitution is needed for repackaging of MPFP, which requires the addition of sugar, and additives to adjust the water activity, pH, and control of browning reaction. The syrup covers the fruit inside the package and protects against microbial contamination. It should have a pH between 3.0 and 4.1. The tank holding the fruit and syrup prior to repackaging should be maintained at constant room temperature for 3 to 5 days during equilibration.

4.5.6 Optimal utilization of the final product

The final MPF product can be eaten as received or used in bulk for off-season processing, in confectionery, bakery goods, and dairy products, or for preserves, jams, and jellies. Fruit pieces can be utilized for salads, barbecue sauces, pizzas, fruit drink formulations, etc.

4.6 Quality control

4.6.1 Recommended microbiological tests

Several microbiological tests should be implemented in the processing area according to the Good Manufacturing Practice (GMP). Microbiological tests also apply to working personnel who manipulate and prepare the fruit products.

Total aerobic counts (TAC): TAC is performed in petri dishes with standard plate count agar (SPCA). These are plated with a spread from the hair, fingerprints, shoe soles, work tables, utensils, and skin of workers with the aid of a wet cue tip, which has been impregnated with a sterile peptone solution (1% v/v). The impregnated cue tip is passed through the desired area being controlled, then spread onto the agar surface in the petri dish. The plates are incubated at 35-37°C ± 2°C for 18 to 24 hours.

Mould and Yeast counts (MYC): To count mould and yeast cells, plates with potato-dextrose agar are plated with the same infected areas described above and incubated for 5-7 days at 25-30°C ± 2°C.

Microbial tests, such as those described above, are also performed on raw fruit to count initial populations, and on the finished product to determine the number of surviving organisms after a combined treatment application.

Knowledge of the combined effect of the preservation factors used for high moisture fruit products (HMFP) on the growth and survival of certain key microorganisms that may pose risks to the quality and safety of HMFP is of great interest in the design of this technology. The major microorganisms of concern in HMFP are primarily moulds and yeasts, due to the high carbohydrate content present in the moisture associated with these products.

4.6.2 Nutritional changes

Very small changes in the nutritional characteristics of MPF are experienced during processing and storage, due to the mild heat treatment applied (compared to thermally processed fruit products). Blanching does not affect the nutritional properties, but it does inactivate the enzymes and provide some reduction of indigenous flora.

4.6.3 Changes in sensory attributes and acceptability

Changes in flavour, texture, odour, and colour have not been reported in high moisture minimally processed fruit products (HMPFP), such as papaya, peach, pineapple, and mango. In general, the average scores presented in Table 4.4 correspond to products that have good acceptability.

Table 4.4 Sensory characteristics of shelf life stable high moisture papaya, peach, pineapple, and mango.

Attribute	Average score
Flavour	6.65-7.70
Odour	5.80-6.80
Texture	6.70-8.07
Colour	6.46-7.10
Overall impression	6.73-7.63

Source: Tapia et al. (1996).

As observed from Table 4.4, texture received the highest scores followed by flavour, colour, and general impression, indicating that combined method technology is a viable alternative in fruit preservation. These parameters are usually judged by using a small trained panel or a larger group of non-trained volunteers. A numerical scale is given for each attribute and the response of each judge is recorded. A scorecard is prepared with a hedonic scale ranging from 0 to 9 points, which is presented to each judge. Nine is the highest score, "like very much", and zero (0) is the lowest score, "dislike very much". Samples are identified with a code number selected at random, as indicated in Figure 4.12.

ACCEPTABILITY TEST

Scorecard **Hedonic Scale**

Judge Name: ______________________ **Date** ______________

Product Name ______________________

Attribute: ______________________

Degree of Preference	**Sample #**	**Sample #**
Like very much		
Like much		
Like moderately		
Slightly like		
Neither like nor dislike		
Slightly dislike		
Dislike moderately		
Dislike much		
Dislike very much		

Comments: ______________________________________

Figure 4.12 Scorecard for sensory evaluation of fruits and vegetables.

CHAPTER 5

PROCEDURE FOR VEGETABLES PRESERVED BY COMBINED METHODS

5.1 Preliminary Operations

Vegetables are subjected to several preliminary operations before processing and after harvesting. As a result of peeling, grating and shredding, produce will change from a relatively stable product with a shelf life of several weeks or months to a perishable one with a shelf life as short as 1-3 days at chill temperatures. The major preliminary operations include:

Washing: Root vegetables are washed first to remove all field dirt and to allow inspection.

Inspection: Vegetables are inspected for quality to comply with consumer demands.

Selection: Vegetables are selected and graded on a basis of firmness, cleanness, size, weight, colour, shape, maturity, mechanical damage, foreign matter, disease, and insects. This operation can be done manually, or by employing a variety of separation machines to separate and discard unfit produce.

Subsequent Operations:
Peeling, cutting and shredding: Some vegetables such as potatoes and carrots require peeling. Ideal peeling is done very gently, by hand with a sharp knife. It has been reported that hand peeling of carrots increases the respiration rate over that of unpeeled carrots, by approximately 15%, whereas abrasion peeling almost doubles the respiration rate compared to hand peeled carrots (Ahvenainen, 1996). Carborundum abrasion peeled potatoes must be treated with a browning inhibitor, whereas washing is enough for hand peeled potatoes (Alzamora et al., 2000). These authors proposed the following guidelines for prepeeled and sliced potatoes:

Processing temperature	4-5°C
Raw material	A suitable variety or raw material lot should be selected using a rapid storage test of a prepared sample at room temperature. Attention must be focused on browning.
Pretreatment	Careful washing with good quality water before peeling is required. Damaged and contaminated parts, as well as spoiled potatoes, must be removed.
Peeling	1) One-stage peeling: knife machine. 2) Two-stage peeling: slight carborundum peeling first, followed by knife peeling.

Washing	Washing is done immediately after peeling. The temperature and amount of washing water should be 4-5°C and 3 L/kg potato, respectively. Washing time: 1 min. Observation: microbiological quality of washing water must be excellent. In washing water, for sliced potatoes in particular, it is preferable to use citric acid with ascorbic acid (maximum concentration of both, 0.5%), combined with calcium chloride, sodium benzoate, or 4-hexyl resorcinol to prevent browning.
Slicing	Slicing should be done immediately after washing using a sharp knife.
Straining off	Loose water should be strained off through a colander.
Packaging	Packaging is done immediately after washing in vacuum or in gas mixture of 20% CO_2 + 80% N_2. The head space volume of a package is 2 L/kg of potatoes. Suitable oxygen permeability of packaging materials: 70 cm^3/cm^2, 24 hr, 101.3 kPa, 23°C, 0% RH (80 μm nylon-polyethylene).
Storage	Preferably in dark at 4-5°C.
Other remarks	Good manufacturing practices (GMP) must be followed (hygiene, low temperatures, and disinfection).
Shelf life	Shelf life of prepeeled potatoes is 7-8 days at 5°C. Due to browning, sliced potato has very poor stability; the shelf life is only 3-4 days at 5°C.

Second washing and drying: Commonly, a second washing is needed after peeling and/or cutting (Alzamora et al., 2000; Ahvenainen, 1996). For instance, Chinese cabbage and white cabbage should be washed after shredding, whereas carrots should be washed before grating (Alzamora et al., 2000). Washing after peeling and cutting removes microbes and tissue fluid, thus reducing microbial growth and enzymatic oxidation during subsequent storage. Washing fruits and vegetables in flowing or carbonated water is more preferable than dipping the product into a tank of water. The microbiological and sensory quality of the washing water must be good and its temperature low, preferably < 5°C. The recommended quantity of water used is 5-10 L/kg for produce before peeling/cutting and 3 L/kg after peeling/cutting (Alzamora et al., 2000).

Preservatives can be used in the washing water to reduce microbial load and to retard enzymatic activity, thus improving the shelf life and sensory quality of produce. The recommended dosage for chemical preservatives in washing water is 100-200 mg/L of chlorine or citric acid (Alzamora et al., 2000). These levels are effective in the washing water before, after, or during cutting to extend the shelf life. However, when chlorine is used, vegetable materials require a subsequent rinse to reduce the chlorine concentration to the level of drinking water and to improve the sensory shelf life. The effectiveness of chlorine should be improved by using low pH, high temperature, pure water, and correct contact time (Alzamora et al., 2000). The optimum contact time for chlorine is 12-13 s, if the chlorine concentration is 70 mg/L (Ahvenainen, 1996). According to Ahvenainen (1996), chlorine compounds are effective in inactivating microorganisms in solutions and on equipment, and in reducing the aerobic microorganism count (i.e., in some leafy vegetables such as lettuce, but not necessarily in root vegetables). Chlorine compounds are not very effective at inhibiting the growth of *Listeria monocytogenes* in shredded lettuce or Chinese cabbage.

Another disadvantage of chlorine is that some food constituents may react with chlorine to form potentially toxic reactive products. Thus, the safety of chlorine use for food or water treatment has been questioned, and future regulatory restrictions may require the development of alternatives. Some proposed alternatives are chlorine dioxide, ozone (0_3), trisodium phosphate, and hydrogen peroxide (Alzamora et al., 2000). The use of hydrogen peroxide (H_2O_2) as an alternative to chlorine for disinfecting freshly cut fruits and vegetables shows some promise. H_2O_2 vapour treatment appears to reduce the microbial population on freshly cut vegetables such as cucumbers, green bell peppers, and zucchini. Moreover, H_2O_2 vapour treatment extends the shelf life of vegetable products without leaving significant residues or causing loss of quality. However, more research is needed to optimize H_2O_2 treatments with regard to efficacy in delaying the growth of spoilage bacteria in a wide variety of vegetable products.

According to Alzamora et al. (2000), the following processing guidelines for shredded Chinese cabbage and white cabbage should be followed:

Processing temperature	0-5°C
Raw material	A suitable variety or raw material lot should be selected using a rapid storage test on a prepared sample at room temperature.
Pretreatment	Outer contaminated leaves and damaged parts, as well as stem and spoiled cabbage, must be removed.
Shredding	Shelf life of shredded cabbage: the finer the shredding grade, the shorter the shelf life. The optimum shredding thickness is about 5 mm.
Washing	Done immediately after shredding in two stages. Temperature and amount of washing water: 0-5°C and 3 L of water/kg of cabbage. Washing time: 1 min. N.B: microbiological quality of washing water must be excellent. Stage 1: • Washing with water containing 0.01% active chlorine or 0.5% citric acid. Stage 2: • Washing with plain water (rinsing).
Centrifugation	Done immediately after washing. The centrifugation rate and time must be selected so that centrifugation removes only loose water and does not break vegetable cells.
Packaging	Done immediately after centrifugation. Typical packaging gas is air with a headspace volume of 2 L/kg for cabbage. Suitable permeability of oxygen for the packaging material is between 1,200 (e.g., oriented polypropylene) and 5,800, preferably 5,200-5,800 (i.e., oriented polypropylene-ethylene vinyl acetate) cm^3/cm^2, 24 hr, 101.3 kPa, 23°C, 0% RH. For white cabbage, perforation (one microhole $150/cm^3$) can be used. The diameter of a microhole is 0.4 mm.

Storage	Preferably in dark at 5°C.
Other remarks	Good manufacturing practices (GMP) must be followed (hygiene, low temperature, and disinfection).
Shelf life	Seven (7) days for Chinese cabbage and 3-4 days for white cabbage at 5°C.

Waxing: During washing, fresh vegetables (also fruits) lose part of their outer layer of wax, which protects against humidity loss. As a result, waxing is re-established after washing with an artificial layer of wax that has adequate thickness and consistency to improve appearance and to reduce the loss of water.

Classification: The main objective of this operation is to attain a uniform product for the market. Fresh vegetables are classified by size, weight, or degree of maturity. Classification by size can be done manually in small packing houses with trained personnel. In mechanized packing houses, operations are carried out with perforated belts, divergent belts or cylinders, and sieving. Sorting by mass is usually done electronically but some manually operated machines can classify different weights by a tipping mechanism. Classification by degree of maturity can be done using colour charts or by optical methods.

Labelling: Commercial fresh vegetables (as well as fruits) can be labelled individually with automatic adhesive stickers to identify the product brand, farmer, or retailer. This is extremely important when exporting produce to other countries.

Premarketing Operations:

Packaging: This operation is one of the most critical in the marketing of vegetables, and involves putting a number of required units in the appropriate package according to weight. Generally, for exporting fresh fruits, corrugated fibreboard boxes of variable capacity are employed.

The most common packaging technique for prepared raw vegetables and fruits is modified atmosphere packaging (MAP). The basic principle of MAP is that a modified atmosphere can be created passively by using specified permeable packaging materials or by actively using a specified gas mixture in combination with such materials.

Storage: The packaged fresh product can be stored at ambient or refrigeration temperature until it is shipped to the overseas market.

5.2 Combined optional treatments

5.2.1 Irradiation

Most studies conducted on irradiation of vegetables (also on fruits) have been targeted to alter ripening and to control post-harvest pathogens and disinfectants. Several countries are exploring alternative methods suitable for the control of human pathogens in fruits and vegetables, and ionization radiation could be one such alternative. It has been demonstrated in literature that application of ionizing radiation (irradiation of foods) is an effective technology for controlling spoilage microorganisms and for increasing the shelf life of strawberries, lettuce, sweet onions, and carrots. Although extensive studies exist on the control of pathogens in meat and poultry products with irradiation, very few studies exist on the value of ionizing radiation in eliminating food borne pathogens in fruit juice, fruits and vegetables, such as lettuce and sprouts (including the seeds to grow sprouts) (Thayer and Rajkowski, 1999). Pathogens in these foods when eaten raw have not generally been controlled in most parts of the world, although this could often be accomplished by combined methods, such as controlled atmosphere packaging (MAP) and ionizing radiation. Thayer and Rajkowski (1999) presented a review on the different dosages of ionizing radiation applied to vegetables to control spoilage and pathogens. In general, most vegetables can withstand irradiation dosages up to a maximum of 2.25 kGy; higher doses can, however, interfere with the organoleptic properties of food products. Combining irradiation with temperature control and gaseous environment, along with adequate processing conditions, is one of the most effective approaches to vegetable preservation.

The moisture content in foods and the surrounding environment during treatment influence the sensitivity of microorganisms to irradiation. For example, high environmental relative humidity and high water content in foods reduce the effectiveness of irradiation; therefore, control of these parameters during irradiation treatments could extend the shelf life and quality of irradiated vegetables. Recently, disinfections of vegetables with chlorinated water have been replaced with irradiation treatments. Treatment of shredded carrots with irradiation at 2 kGy inhibited the growth of aerobic and lactic acid bacteria, in which case sensory analysis panellists preferred the irradiated vegetables (Thayer and Rajkowski, 1999).

5.2.2 Refrigeration

Refrigeration of vegetables can halt the growth of certain pathogen and spoilage microorganisms but will not eliminate them. (The reduction of temperature increases the lag time and decreases the growth of microorganisms). It is generally recognized that maintaining foods at 5°C is sufficient to prevent the growth of most common food-borne pathogens. However, some emerging psychrotrophic pathogens such as *Listeria monocytogenes, Yersinia enterocolitica, Clostridum botulinum* types A and E, *Aeromonas hydrophyla, (enterotoxigenic), and E. coli* are able to multiply slowly in refrigerated foods. Therefore, refrigeration cannot be solely relied upon to maintain the safety of high moisture foods (HM). Considering the increased popularity of MPR foods, this issue has great significance, because refrigeration may be the only hurdle in the preservation of such products. And since psychrotrophic pathogens might eventually prevail, additional factors in the preservation system are needed for safety assurance.

In the conventional refrigeration storage environment, three important factors must be controlled: temperature, relative humidity, and air movement.

Temperature: The system should always be able to meet the demands placed upon it and controlled automatically by the use of thermocouples, pressure valves, etc.

Relative humidity (RH) should be kept high in a refrigerated storage room by controlling the refrigerant temperature. High RH prevents water loss affects texture, freshness, colour appearance and overall quality of food products.

Air movement in the refrigeration environment must be sufficient to remove respiration heat, gases, and the heat penetrating through the door, junctions, and structure of the refrigeration room. However, excessive air movement can cause food dehydration. Air circulation must be uniform throughout the room. Packages must be correctly stacked to achieve good air circulation. Optimum temperatures, RH levels, and expected shelf life of stored horticultural products are described in Table 5.1.

5.2.3 Modified atmospheres

A modified atmosphere (MA) implies removal of, or the addition of gases, resulting in an atmospheric composition different from the one normally existing in air. For example, the N_2 and CO_2 levels may be higher, and the O_2 levels lower than those found in a normal gaseous atmosphere (78% N_2, 21% O_2, and 0.03% CO_2). In this type of storage, the CO_2 and O_2 levels are not controlled under specified conditions.

Appropriate use of MA can supplement refrigerated storage for some products, which could translate into considerable reduction in post-harvest loss. Major benefits can be obtained from its use:

- Reduction in senescence associated with biochemical changes, such as reduction in respiration rate and ethylene production, softening, and compositional changes in fresh produce.
- Decreased sensitivity of fruit to ethylene action at levels of O_2 and CO_2 below 8% and 1%, respectively.
- Can relieve some physiological disorders such as cooling damage in a variety of products.
- Can have a direct or indirect effect on post-harvest pathogens and insect control.

Some disadvantages of MA include:

- Initiation of physiological damage, such as black spots in potatoes.
- Irregular maturation of certain fruits, such as bananas and tomatoes ($O_2 < 2\%$ and $CO_2 > 5\%$).
- Abnormal development of flavours and odours at low O_2 concentrations (anaerobic conditions).
- Increase susceptibility to diseases.
- Stimulation of sprouting and delay of epidermis development in roots and tubers (e.g., potatoes).

Table 5.1. Optimum refrigeration temperature, relative humidity, and shelf life for horticultural products.

Vegetable	Optimum storage conditions Temperature (C°)	Relative Humidity (%)	Expected Shelf life storage
Onions	1 to 2	70 to 75	4-5 mo
Garlic	0	70 to 75	6-8 mo
Beets	0	90-95	1-3 mo
Carrots	0	90-95	4-5 mo
Cabbage	0	98	3-6 mo
Lettuce	0	90-95	2-3 mo
Broccoli	0	90-95	7-10 days
Cauliflower	0	85-90	2-3 weeks
Celery	0	90-95	2-3 mo
Sweet corn	0	85-90	4-8 days
Tomato	12.5-13	85-90	2 weeks
Green pepper	10	95	2 weeks
Chili pepper	10	95	2 weeks
Egg plant	10 to 12	95	3 weeks
Cucumber	10 to 13	95	10-14 days

(From: Flores Gutierrez, A.A, 2000)

5.2.4 Pickling

Vegetables can be macerated in a brine solution for pickling, which preserves the product for a long time. The high concentration of salts in the brine inhibits the growth of microorganisms that decompose and change the flavour, colour, and texture of vegetables. Vegetables can be maintained under maceration with a salt concentration of 6 to 10% during the first ten days of the pickling process. Then, the salt concentration is gradually increased to 16% for six weeks.

Under these conditions, vegetables can be kept in barrels for long periods until final processing. This involves washing the vegetables with water to release large amounts of salt (as much as possible), and packaging the product into glass jars with 5% vinegar and 3% salt. An alternative method is to precook vegetables at 80 to 90°C for 2 to 10 min. Then, packaging is performed using a blend of 3% salt, 6% vinegar, and 5% sucrose.

5.2.5 Fermentation

Vegetables can also be preserved by a fermentation process. During fermentation of raw vegetables, lactic acid bacteria develop, transforming the natural sugars present and the added sugar into acid. In general, a low salt concentration of 3-5% is used to prevent the growth of spoilage bacteria, while lactic acid bacteria are under development. The characteristic flavour and texture of fermented vegetables is produced by the action of lactic acid bacteria. Vegetables must be kept submersed in the liquid to prevent contact with air, which can cause decomposition, due to action of yeasts and moulds. During the fermentation process (2 to 3 weeks), the salt becomes diluted due to water drained from the vegetables, therefore salt must be frequently added to maintain the concentration at 3 to 5%. The pickled vegetables are washed with water and packaged into glass jars containing a solution of 3% salt and 5% vinegar. Vegetables can be pasteurized (or the liquid heated) and packages hot filled.

A typical formulation and application:

Lactic acid fermentation occurs when small amounts of salt are used. This allows bacteria to convert the sugar in vegetables to lactic acid; the acid mixed with salt inhibits the growth of other microorganisms that would cause major damage. This type of fermentation is used to prepare sauerkraut or sour cabbage, and in pickling cucumbers (pickles). Because salting is very softening, -both vegetables and salt are edible, thus preserving nearly all of the nutrients.

A typical application is in the preparation of sauerkraut:

1. Select good, mature cabbages; remove external leaves; wash remaining heads well.
2. With a sharp knife cut the heads into four sections, removing the hearts. Slice two and a half kilos of cabbage into fine strips approximately 2 to 3 cm long.
3. Put above cabbage in pot or plastic container and mix well, adding two tablespoons of salt. Let stand for 15 minutes or more, while preparing another batch of cabbage. The quantity of salt added must be in accordance with the amount of cabbage used for proper fermentation. While the cabbage is in repose, the salt works to reduce the lot size, extract the juice, and soften the cabbage. This will prevent breakage of strips during packaging.
4. The cabbage is packed into clean wide-mouth 4 L glass or plastic jars.
5. Eliminate air bubbles from the cabbage by pressing hard with hand. This allows juices to penetrate the tissues and holes formed between strips. Soft pressing is recommended to avoid breaking the finer strips.
6. Place plastic bag full of water on top of the cabbage to prevent air from penetrating the container and the cabbage. Close the jars tightly. After approximately 24 hours of fermentation, the juices should have completely covered the cabbage. Otherwise, add a brine solution composed of 25 g of salt per L of water until all cabbage strips are covered. The presence of bubbles is an indication that fermentation is in progress. This process lasts from 5 to 6 weeks or until the bubbles disappear from the solution, after which the fermented cabbage is heated in a pot until boiling.

7. Pack cabbage into sterile jars and cover with hot juice, leaving a head space of 2.5 cm below the jar's rim.
8. Place lids on each jar and sterilize the jars in a boiling water bath for 15 and 20 minutes for 0.5 L and 1 L jars, respectively.

5.3 Packaging methods

5.3.1 Plastic containers and bags

The plastic containers used to handle fresh and processed vegetables are tough, easy to handle due to light weight, can be reused, and facilitate the stacking of produce into piles without damaging the product. Initially, the cost of plastic containers and bags is high, but if protected from sun and extreme conditions, they can last for years. In developed countries various plastics are used at all stages of post-harvest packaging and processed fruits and vegetables. Polyvinylchloride (PVC) is used primarily for overwrapping, while polypropylene (PP) and polyethylene (PE), for bags, are the films most widely used for packaging minimally processed products.

Plastic bags are suitable for handling small amounts of vegetable products, and used at supermarkets and retail stores in developed and developing countries.

5.3.2 Vacuum packaging

Vacuum packaging extends the shelf life of vegetables for long periods. This technique relies on withdrawing air from the package with a suctioning machine. Removal of air retards the development of enzymatic reactions and bacterial spoilage. Vacuum packaging and gas flushing establish the modified atmosphere quickly and increase the shelf life and quality of processed products. For example, browning of cut lettuce occurs before a beneficial atmosphere can be established by the product's respiration. In addition to vacuum packing the specifics of handling must be taken into account, especially the time delays and temperature fluctuations.

5.3.3 Modified atmosphere packaging

For some products, such as fast respiring broccoli florets, impermeable barrier films with permeable membrane "patches" to modify the atmosphere through the product's respiration are used. It is not yet agreed as to which films and atmospheres are preferred for minimally processed products.

Modified atmosphere packaging (MAP), can be created passively by using proper permeable packaging materials, or by actively using a specified gas mixture, together with permeable packaging materials. The purpose of this procedure is to create an optimal gas balance inside the package, where the respiration activity of the product is as low as possible; on the other hand, the oxygen concentration is not detrimental to the product. In general, the objective is to have a gas composition of 2-5% CO_2, 2-5% O_2, and the rest nitrogen. One limitation in the design of control atmosphere packaging is in finding good permeable material that will match the respiration rate of the produce; only a few choices are available in the market. Most films do not result in optimal O_2 and CO_2 atmospheres in products with high respiration rates. This problem can be tackled by making micro holes of defined sizes and quantity in the material to prevent anaerobiosis. Another alternative is to combine ethylene vinyl acetate with oriented polypropylene and low-density polyethylene at a specified thickness. These materials have significantly higher permeability than the polyethylene or oriented polypropylene used in the packaging of salads, gas permeability should however be even higher. These materials have good heat-sealing properties and are commercially available.

High O_2 MAP treatment has been found to be particularly effective at inhibiting enzymatic browning, preventing anaerobic fermentation reactions, and inhibiting aerobic and anaerobic bacterial growth. The modified atmospheres that best maintain the quality and storage life of minimally processed products have been found to have an oxygen range of 2 to 8 percent and carbon dioxide concentration of 5 to 15 percent (Cantwell, 2001).

5.4 Transport, storage, and use of vegetables preserved by combined methods

5.4.1 Open vs. refrigerated vehicles

After harvest, open vehicles (trucks, tractors, trains, boats, ships, etc.) are used to transport the product to the packing houses and retail markets. These vehicles are not equipped with refrigeration units and thus the produce decays faster, compared to that in refrigerated transport. If the produce is treated with chemicals or additives after harvest, it can withstand longer distances in open vehicles, without noticeable damage, especially in cases where produce is consumed or processed upon reaching its final destination. Refrigerated vehicles (trucks, trains, ships, airplanes, etc.) contain installed refrigeration units with sufficiently low temperatures to maintain vegetables in a fresh-like state. These types of vehicles are hermetically sealed with insulation material inside the walls of the cave or container, which maintains the cooled product at maximum quality. Vegetables must be classified in order to separate those susceptible to cold temperatures (carrots, potatoes, bananas) and those that are not (tomatoes, peppers, eggplants, cucumbers, etc.). This eliminates the possibility of product damage when cooling at low temperatures during transport. Refrigeration temperatures can vary from 0°C (32°F) to 13°C (55.4°F) and RH from 70 to 95%. Maintaining a high RH in the refrigerated container is very important, as it prevents water loss and degradation in product appearance. This can be accomplished through strict control of the temperature. There is usually little or no environmental humidity control available during transport and marketing. Thus, the packaging must be designed to provide a partial barrier against movement of water vapour from the product. Plastic liners designed with small perforations to allow some gas exchange are one option.

5.4.2 Unloading

Unloading vegetables and fruits from vehicles is a very delicate operation and can be done by hand with a box tipper or with the aid of a forklift. Generally, vegetables and fruits are stacked on pallets to ease the unloading process and to prevent damage to the product. Exported crops arrive at the unloading port in bulk containers are unloaded directly into the storage container with the aid of conveyor belts connected from the vehicle to the container.

At village level a range of head packages and barrows are used to transfer crops from the field. Cushioned surfaces must still be used, however, to protect the crop when unloading.

5.4.3 Storage temperature vs. shelf life

As described in Table 5.1, recommended storage temperatures for vegetables can range from 0°C (32°F) to 13°C (55.4°F) with relative humidity between 70 and 95%; under these conditions shelf life can range from days to months. Controlled or modified atmosphere packaging techniques assist in maintaining adequate temperature control and relative humidity for refrigerated products. These systems can be used during transportation of fresh produce for short or prolonged storage periods. During pre-cooling of some vegetables, high levels of O_2 are utilized for shelf life extension. Recently, injection of CO_2 (%) gas into controlled or modified atmosphere systems to control pathogens was carried out.

5.4.4 Repackaging considerations

Repackaging of vegetables is common when the product has been packaged in large containers, such as sacks, boxes, plastics containers, etc. The repackaging process is often carried out using small trays covered with transparent plastic film, which gives the product an appearance more appealing to consumers. Supermarkets and retail stores display packaged vegetables either on refrigerated shelves or under ambient conditions. Some retail stores and market places use open packages so that consumers can also handle the goods.

5.4.5 Optimal utilization of the final product

Optimal utilization of the final product can vary according to consumer demand. In some cases, the demand is for fresher products. Thus, the optimal utilization of fresh vegetables should be for direct consumption, with perhaps very small quantities remaining for processing or industrial uses. The latter usually occurs with seasonal crops when an abundance of fresh produce floods the market. The produce must meet official regulations concerning product safety and quality whether it is marketed as fresh or processed. Vegetables must be free of foreign matters, chemicals, and microbes that constitute a risk to human health. Therefore, good manufacturing practices (GMP) should be followed during handling, transport, and processing of vegetables for human consumption.

In developing countries, preservation of commodities represents a big problem for small farm crops because of the lack of adequate infrastructure to store harvested products. A novel alternative is to use combined methods technology for preserving large quantities of vegetables without using sophisticated equipment. To implement this preservation technology for stable vegetable products with high moisture content (HMVP), the following considerations should be addressed:

- Technology must be easy to use and be located near production centres.
- Technology must be cheap: does not require the use of sophisticated equipment or machinery, or the use of refrigeration or freezer storage.
- Resulting product must be of high quality: safe and tasty.
- Product must retain fresh-like characteristics.
- Product shelf life should be more than 30 days without refrigeration.
- Product may be commercialized as a final product or kept in storage containers for use as raw material in other processes.

Example of HMVP vegetable preparation:

The procedure used to obtain stable vegetable products was discussed in section 4.3 in the previous chapter and is illustrated in Figure 5.1. In this case, lettuce is selected according to size, weight and degree of maturity. Classification of lettuce is done according to microbiological quality, colour appearance and texture. External leaves are removed by hand and cutting of heads and removal of hearts are accomplished with a sharp knife. Lettuce is sliced into strips approximately 2 to 3 cm long, after which washing of slices is performed with chlorinated water (100-200 g/L) and drying by centrifugation. The finished product is packed into polyethylene-zipped bags and stored at 4-7°C for 2-4 weeks.

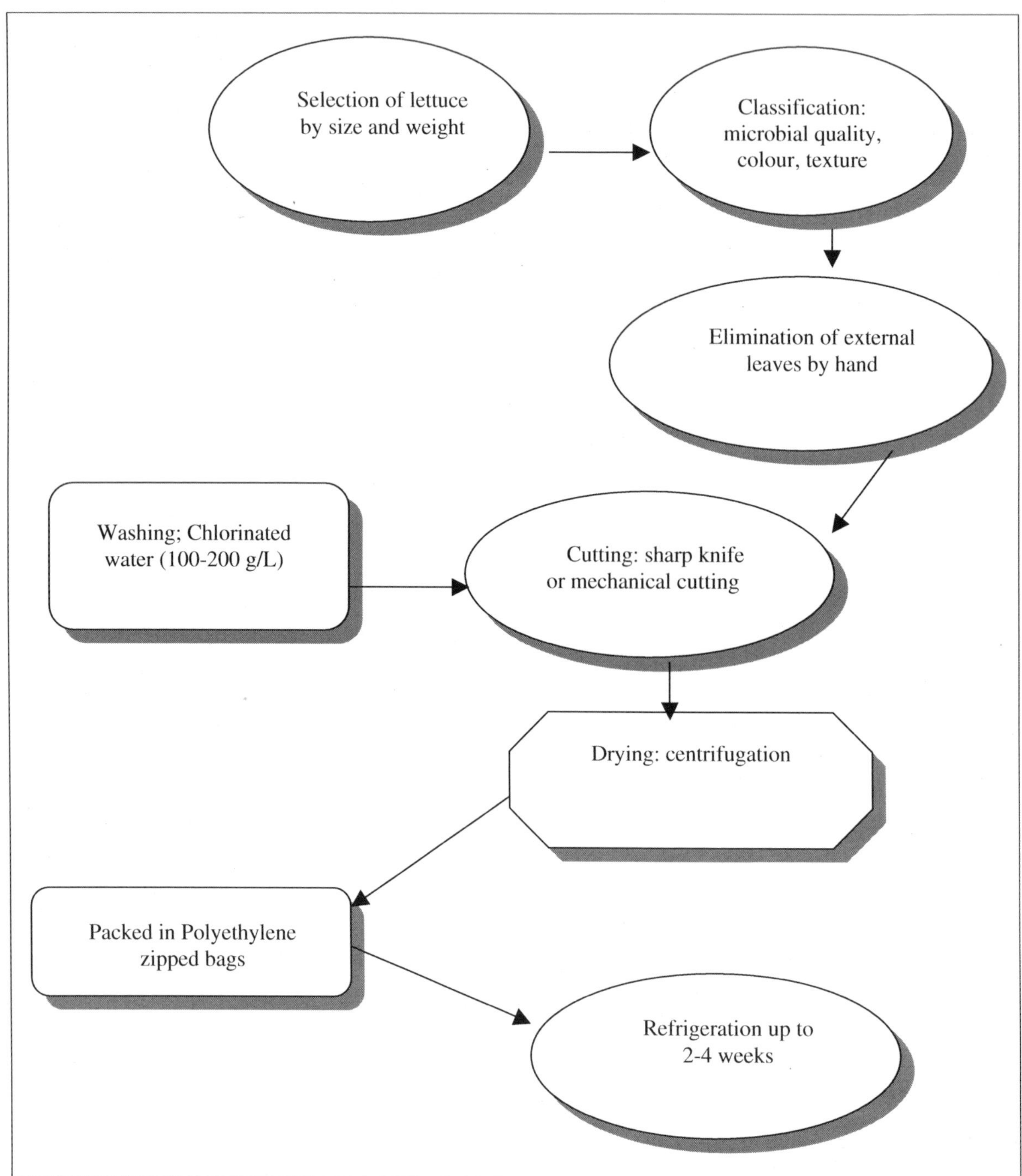

Figure 5.1 Schematic flow diagram to prepare lettuce salad.

5.5 Quality control

5.5.1 Recommended microbial tests

Control tests for microbial invasion in fresh vegetables must be assayed to analyze the growth of spoilage and pathogenic microorganisms. Total aerobic, psychrophile, and coliform bacteria counts are performed in standard plate count agar (SPC) and red violet bilis agar (VRBA). A series of dilutions are made in sterile 0.1% peptone and then pour plated onto SPC and VRB agars; plates for total aerobic and coliform bacteria are incubated at 35-37°C ± 2°C (95-98.6°F ±2°F) for 24/48 hours, and psychrophile bacteria at 7°C ± 2°C (45± 2°F ± 2°F) for 7 days, respectively. However, major contaminants and spoilage organisms in fresh vegetables and fruits or by-products, are moulds and yeasts. These organisms are counted by using potato dextrose agar (PDA) and poured plates, and are incubated at room temperature for 5 to 7 days.

5.5.2 Nutritional changes

Minimally processed vegetables retain nutritional and fresh-like properties because heat is not a major detrimental factor during processing. When using controlled or modified atmosphere packaging in combination with refrigerated storage, prolonged shelf life of vegetable products and retention of vitamins is favoured compared to thermally treated vegetables (e.g., canned vegetables), in which high amounts of nutrients are lost due to severe temperature treatment.

5.5.3 Changes in sensory attributes and acceptability

Since minimally processed vegetables resemble fresh produce, changes in sensory attributes and acceptability are minimized during processing. Thus, flavour, texture, and appearance are retained. Traditional food preservation processes involving high temperature treatments, freezing or dehydration produce an adverse effect, however, on the texture, flavour and aroma of processed food products.

The following factors are critical in maintaining the quality and shelf life of minimally processed products: using the highest quality raw product; reducing mechanical damage before processing; reducing piece size by tearing or by slicing with sharp knives; rinsing cut surfaces to remove released cellular nutrients and to kill microorganisms; centrifugation to the point of complete water removal or even slight desiccation; packaging under a slight vacuum with some addition of CO_2 to retard discoloration; and maintaining product temperature from 1° to 2°C (34° to 36°F) during storage and handling. Temperature maintenance is currently recognized as the most deficient factor in the cool chain (Cantwell, 2001).

Other undesirable sensorial changes are a result of enzymatic activity in raw vegetable products. Two groups of enzymes are responsible for these changes:

Oxidative enzymes such as (Polyphenoloxidase, PPO, and peroxidase) in unprocessed vegetable and fruit products cause browning or other changes in colour. Changes in taste and flavour are caused by lipid oxidation due to the action of the enzyme lipoxygenase.

Hydrolytic enzymes cause softening of vegetable and fruit products (i.e., pectinecterase, and cellulase enzymes); and sweetening of vegetables and fruits by hydrolysis of the starch (amylases). Activity of such enzymes can be prevented by application of thermal treatment, but since the products are minimally processed, the use of heat is not a true option. One has to use other barriers to prevent changes in colour such as anti-browning agents (i.e., ascorbic acid) and anti-oxidizing agents (sulphites) as well as calcium salts to enhance texture firmness of vegetable tissues.

References

Ahvenainen, R. 1996. New approaches in improving the shelf life of minimally processed fruit and vegetables. Trends in Food Science and Technology. 179-197.

Alzamora, S.E., Tapia, M.S., and López-Malo, A. 2000. Minimally Processed Fruits and Vegetables: Fundamental Aspect and Applications. Aspen Pub. Co., Inc., Maryland, US, 277-286.

Alzamora, S.M., Cerruti, P., Guerrero, S., and López-Malo, A., Díaz, R.V. 1995. Minimally processed fruits by combined methods. In: Fundamentals and Applications of Food Preservation by Moisture Control. ISOPOW Practicum II, Barbosa-Cánovas, G. and Welti, J., Eds., Lancaster, PA, Technomic Publishing, 565-602.

Arthey, D. and Ashurst, P.R. 1996. Fruit Processing. Blackie Academic & Professional, London, 142-125.

Barbosa-Cánovas, G.V. and Vega-Mercado, H. 1996. Dehydration of Foods. Chapman & Hall, New York, 53-59.

Barbosa-Cánovas, G.V., Pothakamury, U.R., Palou, E.. and Swanson, B.G. 1998. Conservación No Térmica de Alimentos. Ed. Acribia, S.A., España, 248.

Brown, M.; Bates, R.P.; McGowan, C.; Cornell, J.A. 1992, 1994. Influence of Fruit Maturity on the Hypoglycin A Level in Ackee (Blighia Sapida). Journal of Food Safety. 12 (2), 167-177.

Cantwell, M. 2001. Post-harvest Handling Systems: Minimally Processed Fruit and Vegetables. Vegetable Information. University of California, Vegetable Research and Information Center.

Cook, R.L. 1998. Post-harvest of horticultural crops. Proceedings of the World Conference on Horticulture Research. Acta Horticulture No. 494.

Diaz de Tablante, R.V., Tapia de Daza, M.S., Montenegro, G., and González, I. 1993. High moisture mango and papaya products stabilized by combined methods. CYTED-D, Boletín de Divulgación No. 1, 5-21.

FAO. 1988. Packaging for Fruits, Vegetables and Root Crops. FAO. Rome.

FAO. 1997. Guidelines for Small-Scale Fruit and Vegetable Processors. FAO. Agricultural Service Bulletin 127. Rome.

FAO. 1995a. Small-scale Post-harvest Handling Practices –A Manual for Horticulture Crops. 3rd edition. Series No. 8.

FAO. 1995b. Fruit and Vegetable Processing. FAO Agricultural Services Bulletin 119. Rome.

FAO. 1989. Prevention of Post-Harvest Food Losses Fruits, Vegetables and Root Crops, a Training Manual. Rome.

Flores Gutiérrez., A.A. 2000. Manejo Postcosecha de Frutas y Hortalizas en Venezuela. Experiencias y Recomendaciones. 2nd edit. UNELLEZ, San Carlos, Cojedes, Venezuela, 86-102.

Holdsworth, S.D. 1983. The Preservation of Fruit and Vegetable Food Products. Science in Horticulture Products. Macmillan Press, London, 109-110.

Leistner, L. 1994. Food design by hurdle technology and HACCP. Adalbert Raps Foundation, Kulmbach, Germany.

Leistner, L. 1992. Food preservation by combined methods. Food Research International. 25, 151-158.

Nagy, S. and Shaw, P.E. (1980). Tropical and Subtropical Fruits. Composition, Properties and uses. AVI Publishing, Westport, Connecticut, 127.

Olaeta-Coscorroza, J.A., and Undurraga-Martínez, P. 1995. Estimación del índice de madurez en paltas. In: Harvest and Post-harvest Technologies for Fresh Fruits and Vegetables. Proceedings of the International Conference, Guanajuato, Mexico. February 20-24, 521-425.

Peleg, K. 1985. Produce Handling, Packaging and Distribution. The AVI Publishing Co., Westport, Conn, 20-53.

Proceedings on the World Conferences on Horticulture Research.1996. Global horticultural impact: fruits and vegetables in developed countries. Acta Horticulture 495, 39-66.

Rahman, M.S., Labuza, T.P. Water activity and food preservation. 1999. In: Handbook of Food Preservation, edited by M. Shafiur Rahman. Marcel Dekker, Inc., New York, 339-382.

Ryall, A.L. and Pentzer, W.T. 1979. Handling, Transportation and Storage of Fruits and Vegetables, 320-332.

Ryall, A.L. and Pentzer, W.T. 1982. Handling, Transportation and Storage of Fruits and Vegetables. The AVI Publishing Company, INC. Westport, CT. 2: 98-143.

Sánchez, A., Pérez, J.J., Viloria, Z., Mora, A., and Gutiérrez, G. 1996. Efectos del ácido 2-Cloroetil-fosfórico ("etefon") sobre la composición química en frutos de cambur manzano (Mussa sp (L.), AAB "Silk". Revista Facultad de Agronomía (LUZ). 13, 5-11.

Segre, A. 1998. Global horticultural impact: Fruits and vegetables in developing countries. World Conference on Horticultural Research, Rome, Italy, June 17-20.

Sloan, A.E., Labuza, T.P. 1975. Investigating alternative humectants for use in foods. Food Product Development 9(7), 75-88.

Sofos, J.N. 1989. Sorbate Food Preservatives. CRC Press, Boca Raton, 16-17.

Tapia de Daza, M.S., Alzamora S.M., and Welti, J. 1996. Combination of preservation factors applied to minimally processing of foods. Critical. Reviews in Food Science. and Nutrition. 36(6), 629-659.

Thayer, D.W. and Rajkowski, K.J. 1999. Development in irradiation of fresh fruits and vegetables. Journal Food Technology. 53(11), 62-65.

Thompson, A.K. 1998. Controlled Atmosphere Storage of Fruits and Vegetables. Blackwell Science, Berlin, 14-17.

Thompson, A.K. 1996. Post-harvest Technology of Fruit and Vegetables. Blackwell Science, Berlin.

Welti-Chanes, J., Wesche-Ebeling, P. López-Malo, Vigil, A., and Argaiz-Jamet, A. 2000. Manual de Operaciones para la elaboración de productos hotofrutícolas de alta humedad mediante procesamiento mínimo y métodos combinados. Universidad de las Américas-Puebla.

Wiley, R.C. 1997. Frutas y Hortalizas Mínimamente Procesadas y Refrigeradas. Editorial Acribia, S.A., Zaragoza, España.

Wills, R.B.H., W.B. McGlasson, D. Graham, T.H. Lee, and E.G. Hall. 1989. Post-harvest: An Introduction to the Physiology and Handling of Fruits and Vegetables. Van Nostrand Reinhold, New York.